Chesapeake Invader

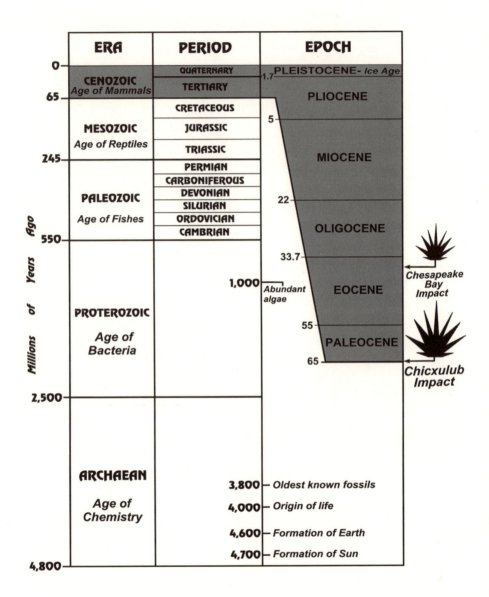

ERA	PERIOD	EPOCH
	QUATERNARY	PLEISTOCENE- Ice Age
CENOZOIC *Age of Mammals*	TERTIARY	PLIOCENE
	CRETACEOUS	
MESOZOIC *Age of Reptiles*	JURASSIC	MIOCENE
	TRIASSIC	
PALEOZOIC *Age of Fishes*	PERMIAN CARBONIFEROUS DEVONIAN SILURIAN ORDOVICIAN CAMBRIAN	OLIGOCENE
		EOCENE
PROTEROZOIC *Age of Bacteria*		PALEOCENE
ARCHAEAN *Age of Chemistry*		

0
1.7
65
5
245
22
550
33.7
1,000 — Abundant algae
55
65
2,500
3,800 — Oldest known fossils
4,000 — Origin of life
4,600 — Formation of Earth
4,700 — Formation of Sun
4,800

Millions of Years Ago

← Chesapeake Bay Impact

← Chicxulub Impact

Frontispiece. Simplified geological timescale used for rock formations, fossils, and events discussed in this book. Epochs are shown only for the Cenozoic Era. The Cenozoic includes the Eocene Epoch, in which the Chesapeake Bay meteorite impact took place.

Chesapeake Invader

DISCOVERING AMERICA'S GIANT METEORITE CRATER

C. Wylie Poag

PRINCETON UNIVERSITY PRESS

PRINCETON, NEW JERSEY

Copyright © 1999 by Princeton University Press
Published by Princeton University Press, 41 William Street,
Princeton, New Jersey 08540
In the United Kingdom: Princeton University Press, Chichester, West Sussex

Library of Congress Cataloging-in-Publication Data

Poag, C. Wylie.
Chesapeake invader : discovering America's giant meteorite crater
/ C. Wylie Poag.
p. cm.
Includes bibliographical references and index.

ISBN 0-691-00919-8 (cl : alk. paper)

1. Cryptoexplosion structures—Chesapeake Bay (Md. and Va.)
2. Geology, Structural—Virginia. I. Title.
QE613.5.C48 P6 1999
551.3'97'0916347—dc21   99-24115 CIP

This book has been composed in Baskerville

The paper used in this publication meets the minimum requirements
of ANSI/NISO Z39.48-1992 (R1997) (*Permanence of Paper*)

http://pup.princeton.edu

Printed in the United States of America

10 9 8 7 6 5 4 3 2 1

To

Dave Folger, who believed,

Gene Shoemaker, who showed the way,

and

Martha, who shared the thrill

Contents

Preface

A SPECTACULAR geological event took place on the United States Atlantic margin about 35 million years ago. It happened in an epoch of geological time known as the late Eocene, a subdivision of the "Age of Mammals." In the late Eocene, sea level was unusually high everywhere on Earth. In eastern North America, the late Eocene shoreline rimmed the flanks of the Appalachian Mountains. Waves from the Eocene Atlantic Ocean broke over the sites now occupied by Richmond, Washington, D.C., Baltimore, and Philadelphia. Late Eocene climate was warmer and more humid than today's Chesapeake summers. Tropical rain forests covered the slopes of the Appalachians, and a broad, lime-covered continental shelf stretched eastward from a narrow coastal plain. Suddenly, with an intense flash of blinding light, that tranquil scene was transformed into a hellish cauldron of mass destruction. From the far reaches of the solar system, a giant meteorite closed in on Earth, traveling sixty times faster than a speeding bullet. It took only a few fiery seconds to brush aside the veneer of atmosphere and ocean and blast deep into the seabed, vaporizing it in a thunderous explosion. A stupendous, supersonic shock wave radiated for thousands of miles in all directions and shook the very foundations of the Appalachians. Searing hypercane winds, laden with white-hot rock debris, roared out of the blast zone followed by towering super tsunamis. Together, they devastated terrestrial animal and plant communities living along the eastern seaboard. Today, the meteorite's legacy is a huge crater, fifty miles across and a mile deep, buried far below younger rock layers and the shallow waters of Chesapeake Bay, Virginia.

This book tells the story of that astral attack and the fifty-year struggle to interpret the clues it left scattered among the rocks under Chesapeake Bay. It is a story of how science works, as its practitioners grapple with incomplete evidence, propose seemingly outrageous hypotheses, and change prevailing concepts of Earth's past history and future evolution. It is a story of collaboration, sometimes intentional, sometimes by chance, among an international cast of industrial, governmental, and academic researchers. It also is a story of serendipity—of being in the right place at the right time with the right stuff.

Acknowledgments

As you read this book, I will introduce most of the key individuals who contributed crucial data, ideas, and hard work that led to the discovery and understanding of this amazing event. Here, I would like to reemphasize my debt to those contributors, and to add thanks to some other special people who played critical roles in this story. Charles Hollister and John Ewing, marine geologists at the Woods Hole Oceanographic Institution, initiated my participation in the Deep Sea Drilling Project in 1970, when they invited me along as a foraminiferal specialist on Leg 11. Yves Lancelot, Chief Scientist for the Deep Sea Drilling Project, selected me as Co-Chief Scientist for Leg 95, during which Jean Thein found the late Eocene tektite layer at Site 612 off New Jersey. Joe Hazel, Chief of the Branch of Paleontology and Stratigraphy, hired me to work for the USGS and sent me to Woods Hole. Dave Folger, close colleague and Chief of the Branch of Atlantic Marine Geology, encouraged me to pursue the meteorite impact hypothesis long before the data were compelling enough to publish. Richie Williams, friend and scientific confidant, urged me to put this story into an accessible public venue. Martha, my patient wife, shared the fervor of discovery during uncounted days when I would come home from the office almost too excited to eat supper. Martha also read and reread all versions of the manuscript. My children, Tracy, Marla, and Graham, served as typical examples of inquisitive non-scientists for whom the Chesapeake Bay story is intended. The manuscript is much improved because of their constructive criticisms. Grover Murray, my father-in-law, and Sally, his wife (both experienced geologists and teachers), provided quite different perspectives for shaping the manuscript. I greatly benefited from the critiques of Don Prothero, an eminent expert on all aspects of Eocene geology, and Peter Ward, whose research and writing have stimulated widespread reevaluation of the biological consequences of ancient meteorite impacts. Jack Repcheck and Kristin

Gager, science editors of Princeton University Press, provided enthusiastic guidance and encouragement throughout the review and publication process. Bill Laznovsky skillfully shepherded me through the crucial tasks of copyediting and proofreading. Ultimately, I must thank all my colleagues and teachers; they convinced me that geologists have a unique responsibility to explain in plain language how Earth's natural systems work, and to point out their implications for mankind.

Chesapeake Invader

CHAPTER ONE

Telltales

JOHN "Cede" Cederstrom couldn't decipher the message. He spent a career gathering clues, but in the end, the crucial key to the puzzle evaded him. Cede (pronounced "seed") worked as a research geologist for the United States Geological Survey (USGS). His main area of research, especially during and immediately following World War II, encompassed the state of Virginia. As America prepared to enter World War II, the myriad naval, army, marine, and air force installations clustered around Chesapeake Bay played a major role in the country's military buildup. The bases expanded rapidly. A principal outcome of this unprecedented growth was a huge demand for new sources of freshwater for drinking and for industrial applications. Finding that water was one of Cede's chief responsibilities. As he searched old well records and drilled new sites, he uncovered telltale signs of an ancient cosmic invader. But the clues were subtle, and he sensed their meaning only vaguely.

The rocks Cederstrom drilled and mapped are part of the Virginia Coastal Plain (fig. 1). This low-lying plain, flanking the Appalachian Mountains, is built up from a thick stack of tabular sedimentary beds. The beds are mainly alternating layers of sand and clay, occasionally interrupted by thin bands of limestone, most of which were deposited when the area was covered by shallow ocean waters. The beds spread out beneath the ground surface of the coastal plain like multiple layers of a gigantic birthday cake, though they thicken gradually and slope gently toward the east (fig. 2). The deepest layers are of Early Cretaceous age, as old as 150 million years. The youngest beds at the top are remnants of the last ice age, or late Pleistocene age, 10–20 thousand years old.

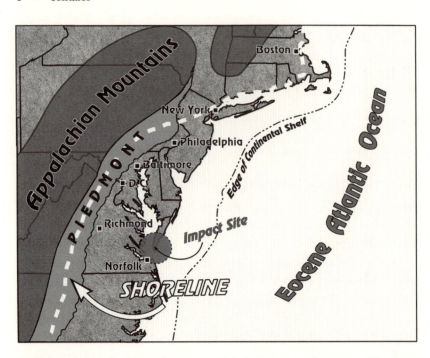

1. Higher sea level and much warmer global climate made the late Eocene a strikingly different world than we know today. The site struck by the Chesapeake Bay meteorite was not a bay in the late Eocene, but a deeper-water location on the outer continental shelf. The shoreline ran along the Piedmont Plateau, west of today's main East Coast cities.

Geologists working in Virginia, including Cederstrom, have organized the coastal-plain beds into a chronological succession of formally named rock units called formations. The oldest formations are on the bottom of the stack, the youngest are at the top. Distinctive features, such as color, mineral composition, or fossil content are used to define each formation. The formation name is taken from a geographic feature near where the formation was first described. For example, the Chickahominy Formation is named for Virginia's Chickahominy River. Likewise, the Potomac, Aquia, and Mattaponi Formations are named after other Virginia rivers. Others, including the Piney Point and Yorktown Formations, take the names of local topographic or cultural features—towns, for example.

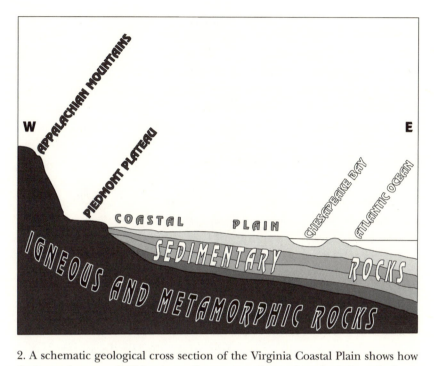

2. A schematic geological cross section of the Virginia Coastal Plain shows how the layered, eastward thickening sedimentary rocks are deposited above the igneous and metamorphic rocks that crop out to the west in the Piedmont Plateau and Appalachian Mountains.

The main body of most sedimentary formations in Virginia is buried beneath the ground surface. The edges of most formations are exposed in cross section, however, wherever they happen to be sliced through by rivers and streams. These surface exposures are called *outcrops*. Some formations, on the other hand, are completely buried, and they can be sampled only by drilling.

Underneath the oldest, most deeply buried sedimentary layers of the Virginia Coastal Plain is a thick layer of very dense *igneous* (granitic) and *metamorphic* (heated and squeezed) rocks, lumped together under the term *crystalline basement*. These buried basement rocks can be traced to the west, where they rise to ground level and become part of the Piedmont Plateau and the Appalachian Mountains.

Many sandy sedimentary beds contain water, stored in tiny, interconnected spaces between sand grains. In reference to their water content, these beds are called *aquifers*. In contrast, clay beds have even smaller pore spaces, which rarely are interconnected. Thus, they act as dense barriers or *confining units*; they prevent or retard the escape of ground water from the aquifers. Fresh ground water can be tapped by drilling through confining units into aquifers, and pumping it out. Geologists refer to the system of aquifers and confining units as *hydrogeologic* units. These hydrogeologic units are given the same names as the formations in which they occur. Part of Cederstrom's job was to identify the specific geological characteristics (thickness, age, minerals, geochemical properties, water quality, and volume) of each individual aquifer and confining unit penetrated by hundreds of drill holes dotting the Virginia Coastal Plain. With this information, he could correlate each bed from well to well and map its distribution over the state.

Geologists can recognize subtle differences between rock formations that are all but invisible to the untrained eye. These differences arise because the environments in which sedimentary layers are deposited change through geologic time at any given locality. Climates heat up and cool down, mountains ascend and erode, sea levels rise and fall, organisms evolve and go extinct. Therefore, each successive sedimentary bed has a different (though sometimes subtly different) mix of minerals, fossils, and pore fluids. Geologists can tell them apart as easily as the FBI can differentiate human fingerprints (fig. 3).

Cederstrom's task was more difficult than usual, because he had inadequate samples of the subsurface formations. The trouble was, the drill bits used to dig the old water wells had to grind or chip their way down through the successive sedimentary layers. As the bit bored downward, it crushed the rocks into fingernail-sized chips, or *drill cuttings*, and scrambled their normal chronological order, mixing older chips together with younger ones. To help retrieve the drill cuttings from the bore hole, drillers pumped viscous *drilling mud* down the hollow drill pipe. This thick mud stream flushed the drill cuttings back to the rig floor through the narrow space between the outside wall of the drill pipe and the

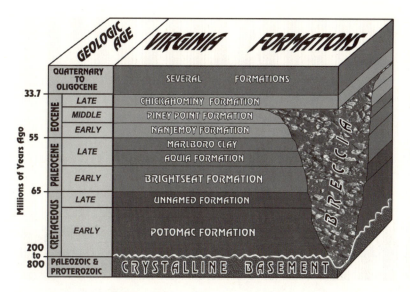

GEOLOGIC AGE		VIRGINIA FORMATIONS	
QUATERNARY TO OLIGOCENE		SEVERAL FORMATIONS	
EOCENE	LATE	CHICKAHOMINY FORMATION	
	MIDDLE	PINEY POINT FORMATION	
	EARLY	NANJEMOY FORMATION	
PALEOCENE	LATE	MARLBORO CLAY	BRECCIA
		AQUIA FORMATION	
	EARLY	BRIGHTSEAT FORMATION	
CRETACEOUS	LATE	UNNAMED FORMATION	
	EARLY	POTOMAC FORMATION	
PALEOZOIC & PROTEROZOIC		CRYSTALLINE BASEMENT	

Millions of Years Ago: 33.7 — 55 — 65 — 200 to 800

3. Succession of sedimentary rock formations that make up the Virginia Coastal Plain. Unconsolidated sedimentary formations are stacked like thick pancakes on top of consolidated igneous and metamorphic formations of the crystalline basement. The oldest (lowest) eight formations were the target rocks into which the Chesapeake Bay meteorite blasted its giant crater. The impact tore rock fragments, or clasts, from each target formation and mixed them together to create a breccia deposit, which filled the crater.

sedimentary wall of the borehole. Such flushing mixed the drill cuttings even more. When the cuttings finally reached the surface, the stream of drilling mud containing them was filtered through a series of sieves. The sieves trapped the cuttings so they could be sampled by the geologist. By careful microscopic examination of a vertical succession of cuttings, Cederstrom attempted to specify the defining characteristics of each individual formation, aquifer, and confining unit. Imagine trying to distinguish all the individual formations penetrated by a 1,000-ft borehole, using only handfuls of rock chips. It's like reconstructing a million-piece jigsaw puzzle. Nevertheless, Cederstrom spent a successful career using this technique to trace and map the beds over hundreds of miles beneath the coastal plain of eastern Virginia (fig. 4).

It would have been much easier and more accurate if Cede had been able to study *cores* rather than drill cuttings. A core is a solid

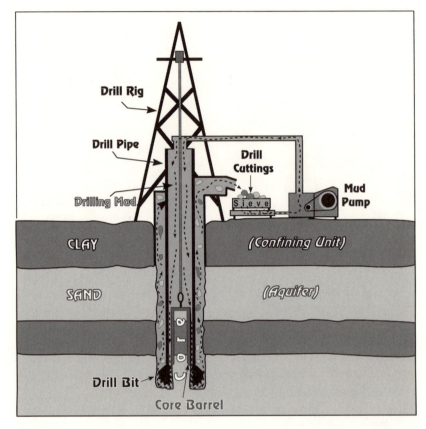

4. Schematic diagram of a drill rig penetrating sandy ground-water aquifers and their confining clay layers. Drilling mud is pumped down the hole to lubricate and cool the rotating drill bit. The circulating mud also flushes drill cuttings from the bottom of the borehole up to collection sieves on the drill floor. To take a core, geologists lower a cylindrical core barrel down the inside of the hollow drill pipe. A tennis-ball-sized hole in the drill bit allows a core of sediment to be pushed up into the core barrel. The core is then retrieved for analysis.

cylinder of rock extracted from the borehole. It allows the geologist to see what the sedimentary layers would have looked like if they had not been ground up by a churning drill bit. A core is the next best thing to an outcrop, because it gives a three-dimensional view of the drilled layers. But, unfortunately for science, it is much cheaper and faster to drill than to core, so it was not customary to take cores. Cederstrom was limited to manually sorting out the drill cuttings, an extremely painstaking task. Only in his mind could he restore them to their original state.

I stand in awe of Cede Cederstrom's interpretive skills. He was truly a magician with drill cuttings. Most of his broad-scale geological deductions are still valid today. I suspect that he might have deciphered those unworldly encryptions in the Virginia rocks, if only he had been able to study cores. As it was, after twenty years of extensive visual and microscopic study, he realized that a particular interval of subsurface strata around Chesapeake Bay was anomalous—downright odd. In a broad swath along the bay's western margin, he found a thick sandy formation that didn't match any of the other beds he had mapped. In each drill hole that contained this anomalous unit, its color, grain size, minerals, and fossils changed rapidly downhole. Mixed in with the usual sand grains were larger rock fragments, or *clasts*. Geologists classify clasts by increasing size into marble-sized *pebbles*, fist-sized *cobbles*, man-sized *boulders*, house-sized *blocks*, and aircraft-carrier-sized *megablocks*. Some clasts were much harder than the unconsolidated sand and clay beds. Their hardness prevented complete destruction by the drill bits, and they stood out among the cuttings as pebbles or cobbles.

The most curious part of this strange deposit was the unusual mixture of microfossils it contained. Cede found abundant remains of microscopic animal species that went extinct 120 million years ago. He was not an expert paleontologist, however, so he called upon his friend Joseph A. Cushman, of Harvard, America's foremost micropaleontologist of that generation. Cushman found that the 120-million-year-old microfossils noted by Cederstrom were mixed together with the remains of species that had lived as recently as 35 million years ago. Intermediate-aged microfossils also were mixed into the same samples. Together, Cederstrom and Cushman determined that the deposit of mixed microfossils was spread over thousands of square miles—an area nearly the size of Connecticut. In recognition of its unusual nature, Cederstrom treated this deposit as a separate formation. He gave it the name Mattaponi Formation, after the Mattaponi River.

But as for how this unusual formation originated, he had no explanation. Remember, he could only study it from small chips. Nowhere did the Mattaponi Formation crop out at ground level, and without cores, he could not determine the original size and shape of the different clasts. He could not see how they were arranged relative to each other in three-dimensional space, nor how the microfossils had been originally distributed within the formation. The natural geometric arrangement of all constituents, which was crucial evidence for interpreting their depositional origin, had been destroyed by the crushing teeth of the drill bit.

Scrambled microfossils were not the only perplexing aspect of the Mattaponi Formation. Instead of containing freshwater, like nearby aquifers of equivalent age, the ground water in the Mattaponi Formation was (and is) very salty. Cede thought that the elevated salinity must have been caused by landward migration of salty waters from Chesapeake Bay. He had no reason to suspect that the briny water might be linked to the unusual composition of the Mattaponi.

The Mattaponi enigma was only one of several subsurface geological puzzles that Cede Cederstrom confronted in southeastern Virginia. For example, he noticed that Eocene formations in the vicinity of Hampton, on the north bank of the James River, were much thicker than Eocene beds on the opposite side of the river, near Norfolk. The thickening took place abruptly, as if a distinct geological barrier ran beneath Hampton Roads, the local name for the lower reach of the James River. Cederstrom interpreted the barrier to be a *fault zone*, a crack in the Earth's crust, along which movement had taken place. He called it the Hampton Roads fault. He believed that the Eocene sediments north of the Hampton Roads fault were thicker because the north block of the fault had descended during deposition, like a slow-moving elevator. This, he reasoned, had created a deep marine basin north of the fault, which could trap a thicker pile of sediments. His mapping indicated that the western margin of the Eocene basin formed an arc, convex to the west, along the western shore of Chesapeake Bay (fig. 5).

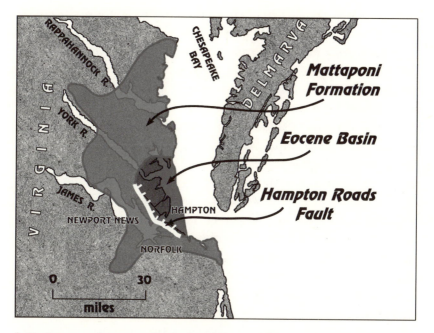

5. During ground-water studies in the 1940s and 1950s, Cede Cederstrom discovered three telltales of an ancient extraterrestrial invader. At that time, however, Cede had no idea that the features were formed 35 million years ago by the Chesapeake Bay meteorite impact. It turns out that the "Eocene Basin" is the western margin of the impact crater, the "Hampton Roads Fault" is part of the crater's outer rim, and the Mattaponi Formation is composed of ejected and scoured rock debris, which fills and surrounds the crater.

Cederstrom's ability to interpret the origin of these features was again limited by the fragmentary nature of his data (drill cuttings). His interpretations were further constrained by traditional concepts of coastal plain evolution that influenced geological thinking during the middle part of this century. He had no reason to connect them with extraterrestrial events.

As we will see, all the geological anomalies Cede Cederstrom identified fifty years ago were telltale evidence of an unearthly intruder. A gigantic meteorite (an asteroid or comet) had blasted out America's largest impact crater in the very region Cederstrom thought he knew so well. We now have evidence that the Hampton Roads fault is the faulted outer rim of the Chesapeake Bay crater, and that the Eocene basin is part of the western sector of the crater itself. Furthermore, the Mattaponi Formation is the deposit of scrambled rock fragments called an *impact breccia*, which fills and surrounds the crater. The briny ground water contained within the Mattaponi is also a product of the ancient impact. We will learn more about that in chapter 12.

During the time of Cederstrom's career, scientists knew very little about meteorite impact structures, and few were trained to recognize the geological traces left in rocks by impacts. The progression of geological knowledge from then to now illustrates one of the fundamental ways in which science works. Scientific understanding continually evolves, much like a lineage of living organisms. We discard outdated hypotheses and propose new ones as we incorporate fresh insights and concepts to answer questions about our natural environment.

Though Cederstrom appears to have been the first to notice nearly all the crucial telltales left by the interplanetary trespasser, he was not the only researcher to misinterpret or ignore fragments of evidence. Several other scientists also missed their chance to identify the crater, either because they did not notice the clues, or because they lacked the proper geological perspective in which to consider them. In 1982, for example, researchers from the USGS's Branch of Atlantic Marine Geology, in Woods Hole, Massachusetts, inadvertently imaged part of the crater. They were carrying out a *seismic reflection survey* in the southern part of Chesapeake Bay and into the mouths of the York and James Rivers.

A seismic reflection survey uses sound waves to image the two-dimensional structure of rock layers beneath the Earth's surface. This kind of remote imaging is especially suitable for scanning geological structures beneath water bodies, because it can be carried out from a ship. The source of the sound waves is a large air compressor, which forces compressed air through a long, thick-walled hose to an *air gun* towed hundreds or thousands of feet behind the ship (fig. 6). The air gun emits a powerful burst of compressed air at carefully timed intervals of a few seconds each. Sound waves from each burst penetrate down through the rock

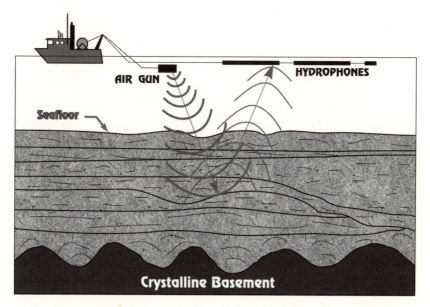

6. During a marine seismic reflection survey, an air gun emits carefully timed blasts of compressed air, which transmit sound waves into the sedimentary and crystalline layers below the seafloor. Part of the wave energy is reflected back to the sea surface each time the sound wave passes a boundary separating layers of different composition. Reflected signals are received by hydrophones, then digitized and processed by a shipboard computer to produce a two-dimensional cross section, or seismic reflection profile, of subsurface structure. Seismic reflection profiles help geologists choose appropriate drill sites to solve particular geological problems or to explore for mineral and energy resources.

layers for a mile or more. Some of the sound energy is reflected back up toward the ship each time a wave passes a boundary between rock layers of strongly differing composition or hardness. This reflected energy is detected by a long string of *hydrophones*, also towed behind the ship. The hydrophones convert the reflected acoustic energy into electronic signals and send them to the ship's computer, which transforms them into digital signals. The digital signals are, in turn, stored on tapes or diskettes, to be further processed onshore. The onshore processing filters out any extraneous sound waves (noise), such as those coming from the ship's engines, and can amplify useful sound frequencies to strengthen weak signals. The final product, a *seismic profile*, is an enhanced, cross-sectional image of the rock layers beneath the sea or bay floor.

On this particular survey in 1982, geophysicist Rusty Tirey and a team of other USGS scientists set out to test some new seismic gear in Chesapeake Bay. Preliminary seismic profiles were printed aboard the ship, but these were difficult to interpret. They were full of reflections caused by extraneous noise, which obscured reflections actually coming from the subsurface beds. Tirey's interpretations would have to wait until he returned to Woods Hole, where more powerful computers could enhance the signals. Once the research party arrived back home, however, the scientists were quickly distracted by other activities, and they postponed the seismic processing. To complicate things further, Rusty was transferred to another duty station shortly after the cruise terminated. As a result, the Chesapeake Bay seismic data were left behind with no one to oversee the necessary processing, so it was never completed.

Ten years later, in the midst of my efforts to document the crater's geometry, USGS geophysicist Debbie Hutchinson remembered those unprocessed Chesapeake Bay seismic data. At her suggestion, we sent the raw data to Warren Agena and Myung Lee, two seismic processing experts in the USGS's Denver research center. To our delight, a profile through the lower part of the James River showed the outer rim of the crater. We could clearly see that the crater rim crossed the James at Hampton Roads, exactly where Cede Cederstrom had placed the Hampton Roads fault. Two other of Tirey's seismic profiles crossed the middle part of the bay; they showed the flat basement surface, which formed the crater floor near its outer periphery. Toward the crater center, the basement surface ascended to a distinct peak. Inward from this peak was a central basin, the deepest part of the crater. The basin was at least half a mile deep, though its total depth could not be determined from the profiles.

Fate had conspired against Rusty Tirey and his colleagues. But even if they had processed the data and noted these features back in 1982, chances are they would not have recognized them as imprints of an ancient meteorite impact. None of his project members had been trained in the specialized subdiscipline of impact geology. The nature of impacts still was poorly known outside a relatively small group of largely self-trained planetary geologists and meteoriticists.

In another case of missed opportunity, USGS geophysicist Peter Popenoe also unintentionally surveyed part of the Chesapeake Bay crater. In 1981, Popenoe imaged the crater rim on a single seismic profile collected just outside the entrance to Chesapeake Bay. The image was not crisp, however, and like Cederstrom and Tirey, Popenoe did not recognize the extraterrestrial implications of the feature.

In 1986, USGS geologists drilling on Virginia's Delmarva Peninsula turned up more telltale evidence of a meteorite strike. But this time the clues appeared in cores. Bob Mixon worked out of the USGS National Center in Reston, Virginia, and spent most of his career exploring subsurface beds of the nearby coastal plain. Bob headed a project to study the nature of the sedimentary beds beneath Virginia's two easternmost counties, Accomack and Northampton. He and project geologist David Powars decided, however, that drill cuttings wouldn't provide the information they sought. In order to optimize interpretations of the subsurface formations, they needed to take cores. They wanted to core continuously from the ground surface to about twelve hundred feet. Instead of yielding centimeter-sized drill cuttings, coring would extract a long, three-dimensional plug from each formation. Cores would measurably improve chances of resolving questions left unanswered by earlier studies of drill cuttings. What was the composition of subsurface formations at this locality? How thick were they? What were their ages? Could they be correlated with the succession of formations on the west side of the bay? Under what environmental conditions did the deposits accumulate? Would some of the sand beds make good aquifers?

Mixon and Powars expected the cores to provide detailed documentation of the vertical succession of minerals, geochemical characteristics, and microfossil assemblages. This geological foresight proved crucial to eventual identification of the impact crater, because the cores contained unequivocal evidence of an extraterrestrial visitor. Once again, however, fate and traditional geological concepts colluded to mislead Mixon and Powars.

Mixon's project set up their truck-mounted drill rig in the coastal marshes near the small town of Exmore, Virginia. This part of the Delmarva Peninsula forms what local citizens call the "Eastern Shore" of Chesapeake Bay. A blazing hot Virginia sun and clouds of mosquitoes from the surrounding murky marshes took their toll on the sweat-soaked drill crew. Nevertheless, coring proceeded rapidly through the loose sands and soft clays near the ground surface. The pace slowed significantly, however, as the hole deepened and the bit encountered harder layers.

The drillers used a *drill string* consisting of 20-ft sections of hollow drill pipe (fig. 7). After every 20 feet of penetration, another section of pipe was screwed on to lengthen the string. The leading end of the drill string contained the drill bit, shaped like a steel doughnut with teeth. Its cutting edge, studded with wedges of carbon steel, could grind through even the hardest granite. The center of the bit, however, was hollow, an empty space about the diameter of a tennis ball. The driller, at the controls of a powerful lifting engine, would lower the rotating bit down through the sediments at a carefully controlled rate. As the bit turned, a solid column of rock extended up through its central opening, into a ten-foot-long core barrel, which rested inside the deepest section of the drill pipe. The driller could insert and retrieve the core barrel by attaching a steel cable to its upper end. Imagine a giant steel syringe extracting a sliver of Earth's sedimentary flesh. At ground level, drillers extruded the core from the barrel, and sent the barrel back down the drill pipe to collect another core. One round-trip of the core barrel could take an hour or more as the hole got deeper.

After retrieving one thousand feet of mostly routine sand and clay deposits, the geologists were astonished by an abrupt change in the physical attributes of the cores. Suddenly, the drillers were pulling out bright, multicolored core segments, which resembled psychedelic barber poles. The dominant constituent of this garish deposit was grayish green sand, whose color came from an abundance of iron-rich *glauconite.* Imbedded within the glauconitic

7. The USGS drilling team extracts cores at Windmill Point, Virginia, on the north bank of the mouth of the Rappahannock River. A 20-foot section of drill pipe is being hoisted into a vertical position so it can be added to the string of pipe already in the borehole. Photograph courtesy of David S. Powars.

sand was a kaleidoscopic array of larger clasts, ranging from dime-sized pebbles to six-foot boulders. The clasts changed rapidly and randomly downcore through nearly every color and hue of the rainbow (fig. 8).

Clast shapes varied from rounded and subspherical to blocky and sharply angular. Many clasts were crushed and shot through with small fractures. Others had been squeezed and deformed plastically, like toothpaste, fresh from the tube. Even chunks of *granite* and *gneiss*, derived from the dense, hard, crystalline basement, were part of this remarkable mixture. Continuous horizontal layers, common to most coastal plain deposits, were essentially absent from this deposit. Instead, the layers were fragmented; individual chunks were rotated, overturned, or scattered in random orientations (fig. 9).

Dave Powars realized that this rocky pastiche must be part of Cederstrom's Mattaponi Formation. He and Mixon were observing it for the first time in its natural three-dimensional state. Powars informally referred to it as the Exmore beds, after the nearby town of Exmore, because most coastal-plain experts no longer accepted the name Mattaponi as a valid formation. That is because of its mixed assemblage of microfossils. Common opinion held that Cederstrom's samples had been mechanically mixed by the drilling process, and did not represent the true nature of the deposit.

Mixon's project cored the Exmore beds at one additional site—across the bay at Windmill Point, near the mouth of the Rappahannock River, approximately thirty miles due west from Exmore. In addition, the USGS scientists collaborated with the Virginia State Water Control Board to core the beds at a third and fourth site. The third site was on the southern tip of the Delmarva Peninsula, near the town of Kiptopeke; the fourth was on the west side of the bay, near Newport News. Scott Bruce, a geohydrologist for the state of Virginia, led the coring team at those two sites.

The thickest cored section of the Exmore beds was at Exmore, where 190 feet of the deposit were penetrated. The base of the Exmore beds was not recovered at any site, however, so its maximum thickness could not be precisely determined. The area between the four core sites was enormous, more than 2,000 square miles, and the Exmore beds seemed to be distributed over the entire area. Combined with the 52 non-cored sites where Cederstrom had reported the Mattaponi, the Exmore beds covered an additional 1,000 square miles—a total area the size of Connecticut.

Nothing remotely resembling these astonishing cores had ever before been observed in the Virginia Coastal Plain, or at any other location along the East Coast. Cederstrom had seen the deposit, but only as drill cuttings. What possibly could have created this strange amalgamation? The geologists were not sure, but they ventured a guess. In their report, published in 1992, Powars, Mixon, and Bruce surmised that the Exmore beds must represent the sedimentary fill of some enormous channel eroded into the ancient sea floor. It might have been emplaced, they reasoned, as a huge gravity-driven flow of rock debris; a submarine avalanche, which

two-foot length of core

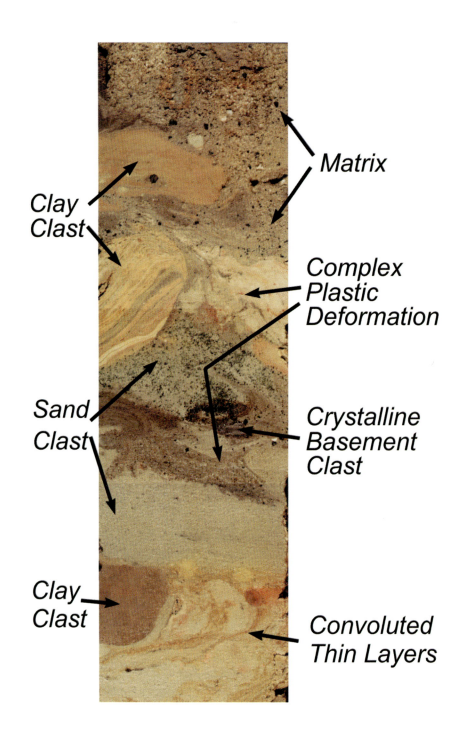

Matrix

Clay
Clast

Complex
Plastic
Deformation

Sand
Clast

Crystalline
Basement
Clast

Clay
Clast

Convoluted
Thin Layers

broke loose from a higher elevation on the ancient continental shelf. If so, it would have been the first such giant debris-flow deposit known to have accumulated on a flat continental shelf. Such deposits, known as *debriites*, usually accumulate along the steep flanks of mountains, or in thousands of feet of water on continental slopes and abyssal plains. The steepness of the land surface or sea floor in these locations helps gravity to initiate and drive the flow. Mixon and Powars realized that theirs was not a geologically satisfying interpretation, so they sought additional analyses.

CHAPTER TWO

Microfossil Magic

ONE of the most important things a geologist needs to know about any rock formation is its age. This knowledge allows him or her to correlate the dated deposit with other beds of the same age in the local area or on other continents. Such correlation leads to a better understanding of how the deposit accumulated and what its ancient environment was like. To determine the geologic age of the Exmore beds, Dave Powars invited three USGS micropaleontologists to analyze the microfossil assemblages contained within the green sandy matrix and in individual clasts. I was one of the three. Each specialist was highly trained to recognize (under the microscope) individual species of a different group of fossilized microorganisms. These microorganisms would have lived in or near the ancient Atlantic Ocean in which the sedimentary beds accumulated. Micropaleontologists have studied the three microfossil groups for a hundred years or more, so the ranges of individual species through geologic time are thoroughly documented. We know when each species evolved, what its ancestors were, when it became most abundant, and when it went extinct or evolved into other species. Thus, by determining which species were present together in a small sample of sediment, we could determine the age of the sediment and assess some fundamental characteristics of the ancient depositional environment, such as whether it had been deposited on land or in the sea. For marine formations, we could determine the approximate depth, salinity, temperature, and oxygen content of the seawater. For nonmarine formations, we could identify such things as the principal plant types, the relative wetness or dryness of the climate, and the seasonal range in temperature. We chose three different microfossil groups, two ma-

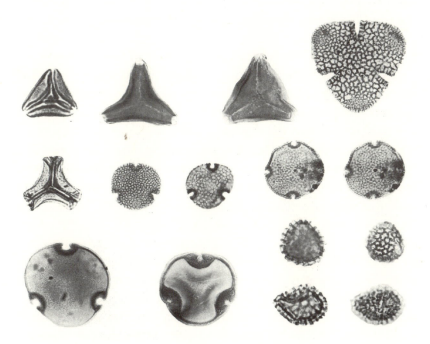

10. Fossilized Eocene pollen grains are terrestrial microfossils. They record the age of the formation which contains them, indicate the types of land plants that existed, and give evidence of the environments where the plants lived. These specimens range in size from 0.001 to 0.002 inch. Photomicrographs courtesy of Norman Frederiksen.

rine and one nonmarine, because we suspected that one or more of the groups would be absent, or poorly represented, in some samples. By using three groups, we also could define a narrower slice of geologic time than was possible with any single group. We completed the preliminary analyses in 1990.

Ron Litwin studied the microscopic remains of land plants, mainly *spores* and *pollen* grains (fig. 10). Both spores and pollen are widely distributed by the wind, and are extremely useful in dating nonmarine deposits, especially those formed in ancient bogs, marshes, lakes, and deltas. Their presence in both nonmarine and shallow marine sediments gives spores and pollen a special role in correlating terrestrial formations with age-equivalent marine formations. Because the distribution of terrestrial vegetation reflects temperature and rainfall patterns, fossil spores and pollen also are excellent clues to the nature of paleoclimates.

Lucy Edwards studied the *dinoflagellates* (literally, whirling whips), a group of microorganisms whose modern representatives are perhaps best known as the cause of toxic red tides and extensive fish kills. Some readers may have heard the recent clamor about the living dinoflagellate species *Pfiesteria piscicida*. This species appears to have caused massive fish kills in Albamarle Sound, North Carolina, and in Chesapeake Bay. *Pfiesteria* is suspected to have debilitated several persons, as well, through seafood poisoning.

Dinoflagellates are one-celled microorganisms with characteristics of both plants and animals, but are usually classified as plants. There is cellulose in the cell wall and chlorophyll in the protoplasm. Living dinoflagellates are major components of oceanic *plankton*. Their vast floating populations are critical links in the marine food chain. In summer blooms, dinoflagellate concentrations can be as great as 100 million individuals per quart of seawater. Because geographic distribution patterns of living dinoflagellates closely reflect oceanic temperature zones, fossil dinoflagellate assemblages can be used to interpret paleotemperatures of ancient oceans.

Most fossils of dinoflagellates, however, do not represent planktonic forms (fig. 11). Instead, they occur as bottom-dwelling cysts. Each cyst is a sort of leathery protective sheath, which forms after the planktonic parent reproduces, or when some environmental stress interrupts the normal dinoflagellate life cycle. Dinoflagellate cysts became abundant in Jurassic seas, 200 million years ago (though they originated 400 million years ago), and their rich fossil record has been used extensively to date marine sediments.

My responsibility was to analyze the *foraminifera*. Foraminifera (or forams, as we informally refer to them) are microscopic, amoeba-like, single cells of protoplasm. Though biologically simple, they construct intricate exterior shells, or *tests*. Foram tests are amazingly variable in composition, shape, internal structure, and

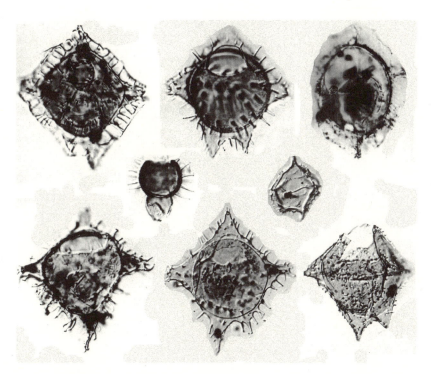

11. Fossilized Eocene dinoflagellate cysts are marine microfossils. They record the age of the formation which contains them, and give clues to the Eocene oceanic environments. These specimens range in size from 0.002 to 0.004 inch. Photomicrographs courtesy of Lucy Edwards.

external ornamentation. Members of most foram species extract calcium, carbon, and oxygen from the seawater in which they live. The forams then combine these elements, secrete the composite as the mineral *calcite*, and use that to build their tests (fig. 12). Representatives of other foram species gather tiny mineral grains from the seabed and plaster them together with organic or mineral-based cement (which the foram excretes) to form an *agglutinated* test. Because of the extreme variability in test construction and exterior ornamentation, a foram specialist can identify different species, genera, and families. The test characteristics also are the keys to evolutionary relationships between different species as they have developed through geologic time. Forams provide some of the best examples of gradual evolutionary transformation of

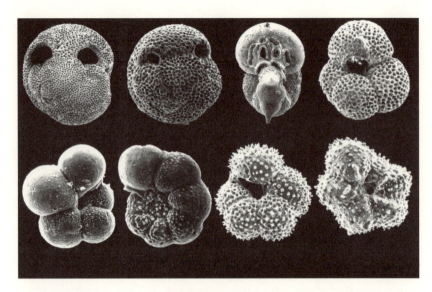

12. Fossilized tests of foraminifera from the Exmore breccia. These marine microfossils record the age of the breccia and of individual clasts within the breccia. They also give evidence of the paleoenvironments in the late Eocene Atlantic Ocean. Each specimen is approximately 0.025 inch in diameter. Scanning electron micrographs produced by Judith A. Commeau.

one species into another species. Micropaleontologists routinely trace subtle changes in test morphology (size, shape, ornamentation) from ancestral species, through intermediate morphologies, into successor species. Evolutionary lineages among foraminifera form the backbone of many prevailing interpretations of the stratigraphic record and the geologic timescale. They are crucial for dating marine rock layers and interpreting the ecosystems, climates, and oceanographic properties of long-lost oceans and seafloors.

Forams have been around for at least 400 million years, and are still present in most environments of modern oceans, bays, and estuaries. They do not, however, live in freshwater or on land. Living foram communities have been studied extensively for more than two hundred years. We know a great deal about the environmental preferences of most modern species and families of forams, because they are still so abundant in marine ecosystems.

Benthic forams thrive by the trillions on and slightly below the seafloor. The test morphology of many benthic species is particularly useful in interpreting environmental conditions on or near the surface of the seabed. Like larger organisms, forams are choosy about their habitats and lifestyles. Some species live only in coral reefs, others prefer muddy mangrove swamps. Still others attach to sand grains or cling to the shells of living clams. Some have developed elaborate attachment stalks; they look like microscopic flying saucers perched on stilts. Other benthic assemblages contain mostly sediment-dwelling species, whose tests provide clues to environmental conditions a few inches below the seafloor. Knowing the ecology and behavior of modern species, including preferred water depth, salinity, temperature, dissolved oxygen content, and food requirements, allows us to determine the paleoenvironments preferred by their evolutionary predecessors.

Equally vast foram populations float among the plankton near the ocean's surface. Some species spend all of their lives suspended in the upper few hundred feet of the water column, whereas other species reside at much greater depths. The distribution of certain planktonic forams is controlled by water temperature; polar and tropical species do not live together. When planktonic forams die, their empty tests collect on the ocean bottom. They are scattered like hieroglyphs across the pages of the geologic record. We can translate these pages to interpret the properties of ancient water masses and paleoenvironmental conditions that existed near the sea surface.

After studying a few samples of each microfossil group, Ron, Lucy, and I agreed that the individual clasts in the Exmore beds had many different ages. The oldest species represented Early Cretaceous populations, 120 million years old; the youngest came from middle Eocene strata, about 40 million years old. We found both marine and nonmarine microfossils within the mixture. Among the foraminifera, both planktonic and benthic species were equally well represented.

Most important, we discovered that the microfossil assemblages were wildly disarranged. There was no orderly upward succession from oldest to youngest species. Instead, some extraordinarily powerful mechanism had ripped the clasts from their original formations, jumbled them together randomly within the sandy matrix, and then redeposited them collectively as a single massive layer over an area the size of Connecticut. Deposition must have been exceedingly rapid and turbulent. Only a few natural forces could have accomplished such a massive disruption and redistribution of strata. How could we determine which one it was?

I decided that a more thorough analysis of the foraminifera might point to the culprit. First, I sampled each individual clast that was hand-sized or larger. Second, I took closely spaced vertical successions of samples from the green sandy matrix. Third, I thoroughly sampled the clay-rich Chickahominy Formation, which overlies the Exmore beds at each core site. This more systematic and detailed examination confirmed that the clasts were fragmented chunks of eight different geological formations known from the Virginia Coastal Plain. In fact, the breccia contained clasts derived from every formation older than the Exmore beds, including crystalline basement, more than a billion years old. Moreover, I could take a single handful of the sandy matrix and find in it microfossils derived from all eight formations. Most important of all, I discovered an additional group of foraminifera, which went unnoticed in the initial study. This group was younger than all the other forams. It clearly showed that the youngest age represented (and thus the age of the Exmore beds) was not middle Eocene after all, but late Eocene. Because the precision and reliability of this age was critical to further interpretations, I called on a friend and colleague, Marie-Pierre Aubry, to provide a second opinion, using a microfossil group that had not yet been analyzed.

Marie is a *nannofossil* specialist at the Woods Hole Oceanographic Institution, on whose campus my office is located. The prefix *nanno-* refers to the fact that these fossils are especially small, not visible to the naked eye. Nannofossils are the limey skeletal remains of a group of one-celled marine algae, which have been a predominant part of oceanic plankton (nannoplankton) for more than 200 million years.

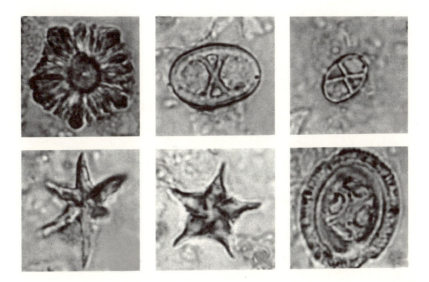

13. Nannofossils from the Exmore breccia. These marine microfossils record the age of the formation which contains them, and indicate paleoenvironmental conditions in the late Eocene Atlantic Ocean. Each specimen is approximately 0.0025 inch in diameter. Photomicrographs courtesy of Marie-Pierre Aubry.

Each tiny cell secretes a crystalline skeleton of minute platelets. The diversity of shapes is remarkable. Circular, cross-shaped, horseshoe-shaped, and star-shaped forms are common (fig. 13). The platelets fall to the seafloor when the cell dies or is eaten. Trillions of platelets pass to the seabed incorporated within the feces of grazing marine organisms. Like the microfossils discussed previously, nannofossil species evolved rapidly through time, and provide a host of extinct species useful in finely calibrating the geological timescale. Nannofossils are so abundant in marine rocks, that samples the size of a pinhead contain thousands of specimens.

Marie examined subsamples of all the Exmore samples I had analyzed for forams. Her age determinations were even more conclusive than mine, because the nannofossils were much more abundant, and their geologic time ranges were shorter than the foram ranges. The nannofossil record confirmed the thorough mixing of clasts from the entire stratigraphic column that existed prior to deposition of the Exmore beds. And the youngest nanno-fossils, were, like the forams, of late Eocene age.

Firm documentation of the late Eocene age turned out to be the magic key that finally unlocked the secret of the Exmore beds and revealed the extraterrestrial origin of Cede Cederstrom's anomalies. But to fully demonstrate the significance of this crucial evidence, we must turn back to the year 1983, and take a trip to the New Jersey Continental Slope on the drillship *Glomar Challenger.*

CHAPTER THREE

Tektites

UNDER a burning August sun, the deck of a drillship stationed off New Jersey is about as hot as the tarmac at Newark Airport. I learned this the hard way in 1983. I was there, as Co-Chief Scientist, to help lead an international party of marine scientists in coring the deep Atlantic seabed. As participants in the National Science Foundation's Deep Sea Drilling Project (DSDP), we used its famous drillship *Glomar Challenger* (fig. 14). This specialized research vessel did not fit the mold of most ocean-going vessels. Three hundred feet long, and forty feet wide, her sleek black hull rode low in the water. On the stern, six decks were crammed with laboratories, workshops, offices, staterooms, storerooms, a library, the galley, and the bridge. What set her apart was the drill derrick; it rose amidships like a miniature Eiffel Tower. Its 15-story spire of steel struts and cables sparkled day and night with phosphorescent green spotlights. With two-inch-thick steel cables strung from the derrick's crown, *Challenger's* mighty draw works could lift a million pounds of drill pipe. She could drill in water so deep that the drill string was more than four miles long!

To drill in such depths, *Challenger* could not anchor. Instead, she maintained position over the drill site by means of computer-controlled *thrusters* combined with the main propulsion system. First, the engineers would drop an acoustic beacon onto the seafloor. Sound waves generated by the beacon were intercepted by four hydrophones attached to the ship's hull. The hydrophones converted the sound to electronic impulses, and sent them to the navigation computer, which calculated the distance between the ship and the beacon. With this information, the computer could, in turn, control the ship's engines to maneuver her over the beacon and keep her there. While drilling, the ship was in constant

14. The drillship *Glomar Challenger*, with its imposing drill derrick, was used by the National Science Foundation to carry out its famous Deep Sea Drilling Project. Photograph courtesy of the Deep Sea Drilling Project.

motion, adjusting to maintain its precise location. Too much drift away from dead center would snap the drill pipe, even though it was nearly as flexible as a spaghetti noodle when dangling a mile or more beneath the ship.

Challenger's coring procedures were much like those used at Exmore and the other Virginia sites, only on a much grander scale. At some sites, the scientific party continuously cored sedimentary formations that were 3,000 feet thick. Altogether, *Challenger* spent fifteen years probing the oceans' geological secrets. She logged more than 375,000 miles on 96 voyages into every ocean and sea, except the ice-covered Arctic. Her scientific parties collected more than 60 miles of core from 1,092 holes at 624 sites. *Glomar Challenger* is no longer in service, but scientific ocean drilling is still supported by the National Science Foundation in collaboration with several foreign governments. The drillship currently in use is *JOIDES Resolution*. JOIDES is an acronym for Joint Oceanographic Institutions for Deep Earth Sampling, an international consor-

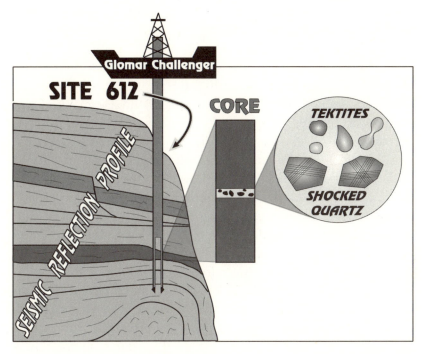

15. During the summer of 1983, scientists of Deep Sea Drilling Project Leg 95 found diagnostic evidence of a late Eocene meteorite impact. Sedimentologist Jean Thein identified tektites and shocked quartz in a core taken from Site 612, 90 miles east of Atlantic City, N.J.

tium of oceanographic institutions, which provides scientific advice and guidance for deep-sea drilling.

During that sultry summer of 1983, we spent seven days and nights coring at a site off the coast of New Jersey, where the water is 4,500 feet deep. We were trying to learn more about the geological framework of the U.S. continental margin, the evolution of its marine life, and the geologic and oceanographic processes that controlled ancient climates and marine environments (fig. 15).

This particular core site, called DSDP Site 612 (it was the 612th core site in the fifteen-year project), is now famous among meteorite-impact specialists, because it produced the first diagnostic evidence that an ancient meteorite had struck the U.S. East Coast. This discovery was a pivotal point for me; it introduced me to the world of meteorite impacts and altered the course of my career.

Without the knowledge gained from Site 612, I probably would not have pieced together the evidence from Chesapeake Bay to reveal America's largest impact crater.

The impact evidence at Site 612 was contained in a chalky layer about eight inches thick. From that layer, Jean Thein, a shipboard sedimentologist from the University of Saarland, Germany, identified *tektites*. Tektites are beads of silica glass, shaped like spheres, tear drops, or dumbbells, which form when a meteorite strikes the Earth. Large meteorite impacts heat the target rocks to hundreds of thousands of degrees Fahrenheit, which vaporizes crustal rocks in the center of the strike zone. Farther away from "ground zero," however, the crust only melts. Droplets of melted silica shoot high into the atmosphere, as if sprayed from a celestial fire hose. If the impact is large enough, the droplets can quickly span the globe, riding atmospheric currents. In the chill atmosphere, the droplets soon cool, harden, and rain back to Earth. Sometimes they deposit a globe-encircling layer a few inches thick. Large tektites are hand-sized or greater. Microscopic droplets are often differentiated as *microtektites* (fig. 16). In the case of Site 612, a rain of microtektites formed a thin layer on the ocean floor off New Jersey. Because no other known geological process produces tektites, they are considered to be unequivocal diagnostic evidence of meteorite impacts.

The tektite layer at Site 612, like the Exmore beds in Virginia, also contained abundant microfossils. The chalk beds above and below the tektite layer also were chock full of microfossils, including forams and nannofossils. The species which Marie-Pierre Aubry and I detected there were the very same species we later identified in the Exmore beds. They indicated that the tektite layer had been deposited 35 million years ago in the late Eocene. And, just like the Exmore beds, the forams and nannofossils in the tektite layer represented a mixture of distinct assemblages from several different geological ages all scrambled together.

That was not all. The tektite layer discovered at DSDP Site 612 contained *shocked quartz*, additional telltale evidence of an impact. The energy released by the enormous shock of a single large meteorite impact far exceeds the force of mankind's entire nuclear arsenal. It would be like the simultaneous detonation of 100,000

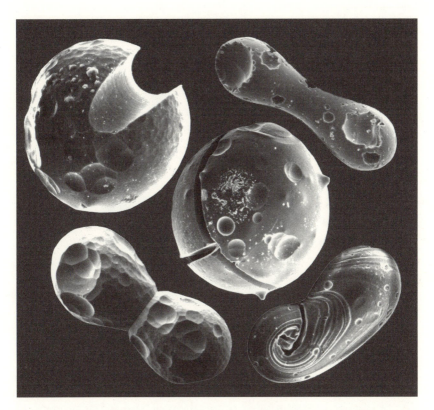

16. Typical shapes of late Eocene microtektites found at Deep Sea Drilling Project Site 612 and elsewhere in the North American tektite strewn field. The pits and grooves are the result of dissolution due to exposure to seawater. The swirled lineations in the lower right-hand specimen are the result of having been twisted while the glass droplet was still molten. Each specimen is approximately 0.04–0.05 inch long. Scanning electron photomicrographs courtesy of Bill P. Glass.

hydrogen bombs. Such tremendous forces alter the crystalline structure of mineral grains in rocks outside the vaporization and melt zones. At most impact sites, the most abundant shock-altered mineral is quartz (silica, or silicon dioxide), so shocked quartz grains are part of the debris ejected from the crater. Shocked grains cannot be distinguished from normal quartz grains with the unaided eye, but are easily identified using a high-powered optical microscope. Unshocked quartz grains are transparent, like glass, but shocked grains exhibit bundles of closely spaced, parallel, dark lines. Usually, several bundles are produced in each grain, and the bundles cross each other at specific angles relative to the optical axis of the quartz crystal.

In addition to shocked quartz, geologist Bill Glass at the University of Delaware has identified two other shock-altered minerals at Site 612; *stishovite* and *coesite*, which are named for their discoverers, Drs. Stishov and Coes. Each of these minerals is a different crystalline form of silica created by progressively increasing pressures derived from the impact shock. Impact shock is measured in units of pressure called *Pascals*, named after the French philosopher and mathematician. It takes 101,325 Pascals to equal the pressure exerted by the Earth's atmosphere at sea level, which is 14.7 pounds per square inch. The number of Pascals exerted by a large meteorite impact is so great that scientists express it in billions of Pascals, or *gigapascals* (GPa). Shocked quartz forms at pressures of 5–50 GPa (5–50 billion Pascals, which equals 50,000–500,000 atmospheres), stishovite forms at 12–45 GPa. A third high-pressure form of silica, *lechatelierite*, forms at 50–100 GPa, and at pressures greater than 100 GPa, silica vaporizes. For comparison, the pressure at the center of the Earth is approximately 400 GPa. Based on this progression, geologists can use the relative abundance of the different silica minerals as a "shock barometer" to estimate the maximum impact pressures transmitted to the target rocks.

In studies published in 1987, Bill Glass and Christian Koeberl (a geochemist at the University of Vienna, Austria) suggested that the Site 612 tektites were part of what they called the *North American tektite strewn field* (fig. 17). North American tektites are strewn over an area larger than 3.6 million square miles, stretching from Barbados, through the Caribbean Sea, to Cuba, into the Gulf of Mexico, and to Texas and Georgia. All were deposited 35 million years ago during the late Eocene, and have similar geochemical compositions. The relatively large size of the Site 612 tektites, their geochemical composition, and the 8-inch thickness of the tektite layer were clues to the general location and size of their source crater. Published estimates put the crater's location within a few hundred miles of DSDP Site 612. The geochemical constituents

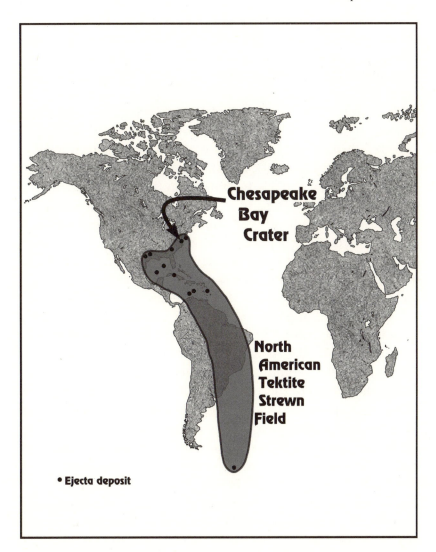

Chesapeake
Bay
Crater

North
American
Tektite
Strewn
Field

• Ejecta deposit

17. The North American tektite strewn field is composed of microscopic to hand-sized minerals and melt droplets ejected from a meteorite crater, presumably that of the Chesapeake Bay impact. Most sample sites are in the northern hemisphere, but a single site in the Southern Ocean also may be part of the strewn field.

of the tektites and other ejecta at Site 612 indicated derivation from rocks with the same general composition as the Appalachian Mountains, or from sediments eroded from Appalachian rocks. Researchers estimated that the crater should be 8–30 miles in diameter.

I was intrigued by the possibility that a large impact crater might be located nearby on the Atlantic Continental Shelf. So I began to search through the large archive of offshore seismic reflection profiles collected by my USGS colleagues in Woods Hole. But at that time, I knew virtually nothing about impact craters. Therefore, I would have to educate myself to recognize what one would look like on a seismic profile. Remember that a seismic profile is essentially a cross-sectional display of whatever feature it crosses. What did an impact crater look like in cross section? Were all craters alike? Could they be confused with other types of structures, such as volcanos? I turned to the published record to find out.

Learning what others have previously written on a subject is a necessary part of problem solving for a scientist, particularly in a historical science like geology. I used the USGS's world-class geological library in Reston to accumulate a large cache of impact information in a few short months. I learned that meteorite specialists have discovered about 150 impact structures on Earth. Of these, only a handful have been preserved well enough and studied thoroughly enough to serve as standard examples. The other rocky bodies of the solar system, on the other hand, contain many hundreds of thousands of superbly preserved impact craters. Of course, they have to be studied remotely, using telescopes, satellites, planetary landers, or rovers. From these diverse data, planetary geologists have distinguished two principal types of craters—simple and complex.

Simple craters are typified by the well-known Meteor Crater, or Barringer Crater, located thirty miles east of Flagstaff, Arizona. Simple craters are essentially small, shallow, bowl-shaped excavations, up to six miles in diameter (fig. 18). The crater rim is a nearly circular or ovate ridge, raised tens of feet above the surrounding ground level. It superficially resembles a volcanic peak. Inside the rim, the excavation is usually partly filled with breccia.

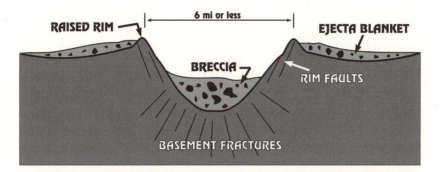

18. Typical geologic and morphologic features of a simple impact crater, as exemplified by Meteor Crater in Arizona. Generalized cross section above; aerial photograph below. Photograph courtesy of USGS.

The breccia deposit has a sandy matrix in which are embedded larger clasts, which were ejected a short distance into the air before falling back into the crater. Other constituents are large blocks that slump from the outer walls of the crater. If the impact takes place in the ocean, additional clasts are produced by vigorous hydraulic erosion as the water column collapses back onto the seafloor after passage of the meteorite. Additional clasts are ripped from the seafloor and the adjacent coastal plain by giant sea waves, or *tsunamis*, generated by the meteorite's splashdown. These clasts are added to the breccia by the backwash of the tsunamis.

The outer rim of an impact crater is a zone of intense fracturing, faulting, and slumping. The steep, inward-facing wall is essentially a single, subcircular, cliff, or fault plane called a *scarp*. In the best-preserved simple craters, the rock layers of the raised rim are folded back upon themselves, so that older rock layers lie on top of younger layers. Geologists call this condition *inverted stratigraphy*. The floor of the crater may be an irregular, blocky surface, honeycombed with faults and fractures.

Complex craters are generally much larger than simple craters, implying significantly larger meteorites. Diameters of complex craters range from six miles to several hundred miles. Also, the diameters are generally a hundred times greater than the depths of excavation (the vertical distance from the crater rim to the crater floor). The outer rim may or may not be raised. Most important, the interior of a complex crater may contain three or four different structural features lacking in simple craters (fig. 19). For example, the outer region of the crater floor resembles a very wide, circular racetrack, and is called the *annular trough*. Next, toward the center of the crater, is a raised, subcircular ring of low peaks, called a *peak ring*. Inside the peak ring is the *inner basin*, the deepest part of the crater floor. In the very center of the inner basin, some complex craters display a single, tall, *central peak*. The floor of complex craters, like that of simple craters, is buried by a thick pile of breccia, formed in the same ways as the breccia in simple craters. Commonly, complex craters initially are excavated deeper into the crystalline basement than simple craters, as the

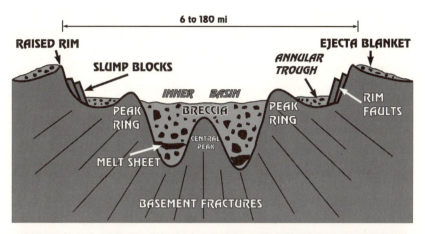

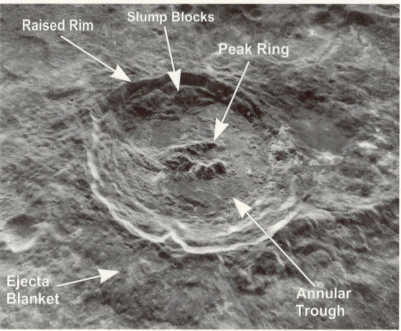

19. Typical geologic and morphologic features of a complex impact crater, as exemplified by King crater on the far side of the Moon. Generalized cross section above; satellite image below. Satellite image courtesy of Ronald Greeley.

result of much greater impact forces. This produces melt zones in the crystalline basement rocks and in the lower part of the breccia. Melt zones are smaller, or may be absent in simple craters. The crystalline basement also is fractured to greater depths (several miles) beneath complex craters. A good example of a complex crater with a peak ring is King crater on the far side of the Moon.

Both simple and complex craters typically are surrounded by an *ejecta blanket*, consisting of rock debris ejected from the crater combined with clasts torn loose from beds nearby the impact site. Composition of the ejecta blanket is similar to that of the breccia inside the crater, but the blanket is much thinner. In most cases studied, ejecta closest to the crater contains only rare specimens of shock-altered minerals or tektites, whereas shocked or melted particles can be quite abundant at greater distances from the crater.

Faults in both simple and complex craters are concentrated around the outer rim and in the crystalline basement, below and adjacent to the crater. Additional faults are created outside the outer rims, especially around complex craters, due to the much more powerful shock waves associated with them. Some shock-generated faults may develop as far as five hundred miles away from the crater rim.

My literature search revealed that only a few seismic profiles of impact craters had been published. Those showed that, in cross section, the major structural elements of well-preserved craters are relatively easy to distinguish from other geological features. The locations, shapes, and sizes of the raised rim, the flat annular trough, the raised peak ring, the deep inner basin, and the up-lifted central peak generally can be identified by their characteristic seismic reflections. The marked contrast between very dense crystalline basement rocks and less dense sedimentary beds overlying them, produces a particularly strong seismic reflection that is easy to identify. The breccia within the crater, on the other hand, is seen on seismic profiles as a thick zone of chaotic seismic reflections. The vast number of irregular clasts within the breccia reflect the seismic acoustical energy in so many different directions that individual reflections interfere with each other and prevent production of a coherent image.

After a few months of studying published reports, I was confident that if a genuine impact crater was present near Site 612, I could recognize it. So I scrutinized the collection of USGS seismic profiles from that area, until I found a place where the subsurface rocks were oddly disrupted. I thought it might be the impact crater I was seeking. It was located near the head of Toms Canyon, a deep submarine channel cut into the New Jersey Continental Slope. Furthermore, it was located only twelve miles from the tektite layer at DSDP Site 612. I soon discovered, however, that I didn't yet have enough evidence to be absolutely sure the Toms Canyon structure was the source of those tektites.

CHAPTER FOUR

Toms Canyon

THE EASTERN continental shelf of the United States is a submarine extension of the emergent coastal plain, composed of the same layer-cake sedimentary beds. Thick, almost horizontal beds of sand, clay, limestone, and chalk dip gently to the east, and thicken slightly as they near the shelf edge. At the shelf edge, the beds crop out in rugged cliffs, eroded in many places by deep submarine canyons. I found two USGS seismic profiles that crossed the New Jersey Continental Shelf within twelve miles of DSDP Site 612. Both showed unusual disruptions of buried late Eocene strata. Unfortunately, however, the profiles were located three miles apart. Without additional profiles to fill in this gap, I could not document the three-dimensional geometry of the feature. All I had was tantalizing, but inconclusive evidence of a possible impact crater. In order to be certain, I needed a much denser grid of seismic profiles. I decided to canvas some of my colleagues in other geologic agencies to seek the location of such a grid.

First I called the Minerals Management Service (MMS), a federal agency that was spun off from the USGS in 1980 to regulate exploration and production of offshore mineral resources. The MMS geologists informed me that a consortium of about fifteen oil companies had collected a closely spaced series of seismic profiles near Site 612 in 1975. A small consulting company had bought the distribution rights to the seismic data. I wrote a letter to the company representative. After months without success by mail, I finally reached him by phone, and learned that his company no longer existed. Their proprietary claim to the seismic profiles had expired, and the data now were archived by Texaco, Inc., the senior partner in the exploration consortium.

I contacted Texaco and asked for copies of selected seismic profiles that crossed the study area off New Jersey. Texaco responded that even though the ten-year proprietary agreement had expired, they could not release the seismic data without consent of the other fourteen partners. They agreed, however, to make an effort to contact the partners and present my request. I worked for Chevron Oil Company during the 1960s, so I understand that requests like mine can be quite irritating to employees who are expected to devote 100 percent of their office time to finding and producing oil and gas. Most companies recognize, however, that a certain amount of scientific cooperation with academia and the federal government is beneficial to all parties.

But running down all the consortium partners was not as easy as it might sound. Some of the partners had been small groups of investors, which had long since been dissolved, and whose legal rights to the data were not clearly known. I am indebted to Texaco geologist Parish Erwin, who worked on the problem for three years until all partners agreed to release the data. I was dismayed by the long delay, but eventually the Texaco seismic profiles arrived in Woods Hole.

The new data were superb. I now had an intersecting grid of 19 seismic reflection profiles, spaced about half a mile apart, right above the buried structure (fig. 20). This grid filled in the area between the two USGS profiles I had initially analyzed. The new profiles clearly showed a crater-like excavation buried by 600 feet of water and 3,000 feet of sedimentary rocks (fig. 21). The center of the structure was about 6 miles west of the head of Toms Canyon, a large submarine channel eroded into the shelf edge, just 16 miles north of Site 612. The crater, which I named Toms Canyon crater, is oblong, 12 miles long, 8 miles wide, and about 1,000 feet from rim to crater floor. The crater has a distinctly raised outer rim, which is highly faulted. The crater floor is irregular, and exhibits elongate ridges and troughs whose long axes trend northeastward. No obvious annular trough, peak ring, inner basin, or central peak can be distinguished. Clearly it is not a complex crater. It resembles a simple crater, except that its outline is

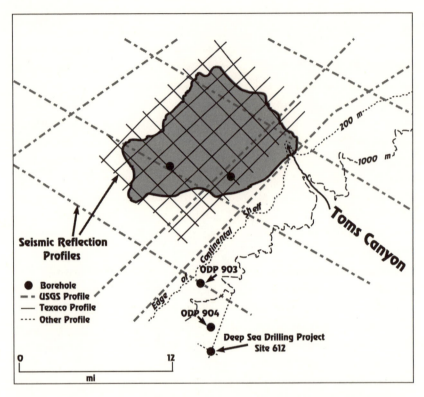

20. Toms Canyon crater, one-fourth the size of the Chesapeake Bay crater, is buried near the head of Toms Canyon, 90 miles east of Atlantic City, N.J. This small crater and ejecta drilled within it and nearby boreholes are the same age as the Chesapeake Bay crater.

neither circular nor ovate. In map view it looks more like an obtuse triangle, whose base is to the northeast.

In order to figure out what would produce such an oddball crater, I was forced to dig into the published literature once again. It didn't take long. Peter Schultz, a planetary geologist at Brown University, had the answer. Schultz had spent much of his career observing and modeling craters formed by low-angle impacts. He fired pellets from the National Aeronautical and Space Administration's (NASA) high-velocity gas gun (at the Ames National Laboratory, in California) to blast small craters into a variety of different materials. In his experiments, pellets that struck at angles

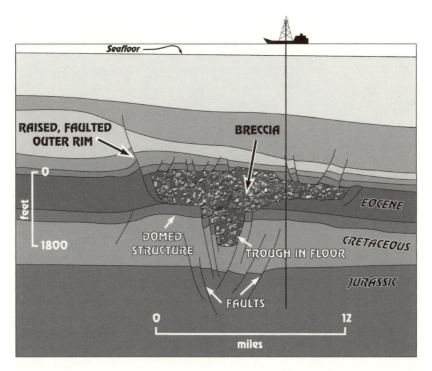

RAISED, FAULTED
OUTER RIM

BRECCIA

Seafloor

feet

0

1800

EOCENE

DOMED
STRUCTURE

TROUGH IN FLOOR

CRETACEOUS

JURASSIC

FAULTS

0   12

miles

21. A schematic cross section shows the breccia-filled Toms Canyon impact crater, the raised outer rim, rim faults, shallow flat floor cut by a deep trough, and domed uplift of the floor.

smaller than 45° to the horizontal, formed ovate craters. In one experiment, he fired a cluster of pellets, which carved elongate ridges and troughs into the floor of the ovate crater, like those at Toms Canyon. The long axes of the ridges and troughs paralleled the trajectories of the impacting pellets (fig. 22).

Schultz's research convinced me that the Toms Canyon crater had been produced by a cluster of small meteorites, which approached the New Jersey shelf at a low angle from the southwest. The cluster would have excavated a crater that was originally ovate, and had a furrowed floor. But how did the ovate crater become triangular? Microfossils again provided an important clue. The microfossils at Site 612 and in the two crater drill holes indicated that the ocean depth at the impact site was around 1,500–2,000 feet in the late Eocene. After the meteorite cluster had passed

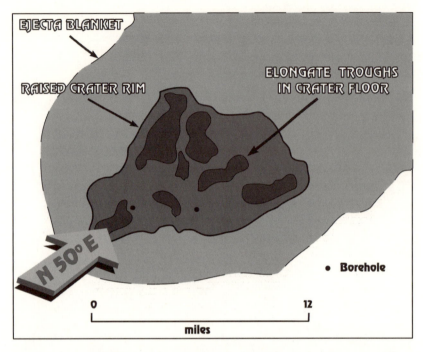

22. The floor of the Toms Canyon crater contains elongate troughs. These troughs are believed to be the result of a cluster of small meteorites, which traveled in a northeast direction, and plunged at a low angle into the deep ocean. After the cluster passed through the deep water column, the water column collapsed back onto the seafloor and washed out the water-saturated crater walls. This formed an irregularly triangular crater.

through a water column that thick, the column would have collapsed to refill the hole, producing tremendous hydraulic pressures on the seafloor. Hydraulic erosion would have washed out huge sections of the crater wall, and destroyed its ovate outline.

The Toms Canyon crater is filled by sedimentary debris, which produces a zone of chaotic seismic reflections. By chance, oil companies had drilled two exploratory wells into that chaotic zone. I obtained samples from each well, courtesy of Dick Benson, a geologist with the Delaware Geological Survey. The samples were drill cuttings, not cores, but they contained more evidence of a meteorite impact. They showed that the zone of chaotic seismic reflections, which I interpreted as impact breccia, contained scrambled assemblages of foraminifera. Species of different geological ages were all mixed up, just like those in the tektite layer at DSDP Site 612 and in the Exmore beds. Most important, the foraminifera contained in all three deposits indicated that the Exmore beds, the Site 612 tektite layer, and the chaotic zone of the Toms Canyon crater, had the same age—late Eocene, about 35 million years old. John Obradovich, a USGS isotope specialist stationed in the Denver Federal Center, confirmed the 35-million-year date by measuring the ratio of two isotopes of Argon gas trapped in some of the tektites at Site 612.

Now you can see how the connection between these three late Eocene events was developing in my mind. The evidence clearly showed a remarkable coincidence of unusual, even spectacular, geologic activity. A small, breccia-filled impact crater was created on the New Jersey continental shelf at the same geological moment in which tektites and shock-altered ejecta accumulated twelve miles to the south at Site 612. Concurrently, two hundred miles to the southwest, some gargantuan force tore apart coastal plain formations over an area the size of Connecticut and redeposited them as an immense breccia formation. Surely all three events must have been generated by the same powerful mechanism—a meteorite impact near Toms Canyon.

If the Exmore beds were the result of a meteorite impact, however, shouldn't they contain shock-altered minerals, or maybe tektites? I asked two more of my USGS colleagues, Larry Poppe (Woods Hole) and Glen Izett (Denver; now retired), who are expert mineralogists, to look for microtektites and shocked quartz in samples of the Exmore beds. Glen was world renowned for his analyses of impact ejecta from the dinosaur-killing meteorite that terminated the Cretaceous Period of geologic time. Sure enough, Larry and Glen found the telltale bundles of intersecting dark stripes in quartz grains from almost every sample of the Exmore matrix (fig. 23). The shocked grains were present in only minute amounts, however. Poppe estimated that less than 0.1 percent of the matrix consisted of shock-altered grains. Neither Glen nor Larry found any tektites. The absence of tektites puzzled us, but nevertheless, we had discovered the smoking gun. There was no

23. Microscopic grains of shocked minerals extracted from Exmore breccia cores taken from the Chesapeake Bay crater: *above,* a single grain of quartz from the matrix; *below,* a thin section of a feldspar grain in a clast of crystalline basement rock. Photomicrographs courtesy of Lawrence J. Poppe and Christian Koeberl.

longer any doubt that the Exmore beds constituted a genuine impact deposit, and I correlated that deposition with the Toms Canyon impact.

Questions still remained, however. For example, if the Toms Canyon impact had generated both the Site 612 ejecta and the Exmore beds, why were the Exmore beds thousands of times thicker than the 612 ejecta, even though they were two hundred miles farther away? Why hadn't anyone found Exmore-like beds in New Jersey, which was closer to the Toms Canyon crater?

CHAPTER FIVE

Super Tsunami

AT that point, I could think of only one way to relate the Exmore beds to the distant Toms Canyon crater—they must be a giant tsunami deposit. In theory, any large meteorite impact into the ocean would initiate a series of gigantic tsunami waves. The maximum estimated height of the waves at the impact site would equal the depth of the water there. The microfossil evidence indicated that the late Eocene water depth at impact was around 1,500–2,000 feet. This meant that the super tsunami would have started out being a quarter to half a mile high, but wave heights would gradually degrade as the tsunami moved away from the impact site. When the waves moved into shallow water near the shoreline, however, they would begin to feel bottom; this would slow down the wave fronts and cause their crests to rise again to *forty times* their open-ocean height. At the shoreline, then, the waves would have been hundreds, possibly even thousands, of feet high. I know some avid windsurfers who would relish challenging a wave like that.

Such a monster wavetrain coming from the Toms Canyon impact would travel at high speed. Modern tsunamis generated by earthquakes travel as fast as 500 miles per hour. Thus wave effects from the Toms Canyon impact would have been felt in southeastern Virginia in less than half an hour. There, the enormous hydraulic power of the churning waters would easily have stripped unconsolidated sedimentary beds from the floor of the continental shelf and from the coastal plain of Virginia, all the way to the Appalachian foothills. The powerful backwash of the waves would transport the scoured-out fragments back into the sea. Eventually, within a few hours or days, the seaward backrush would have piled

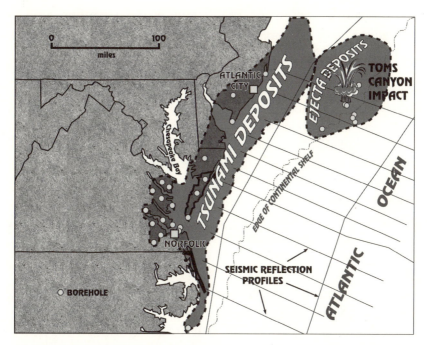

24. I originally inferred that the Exmore beds in Virginia were tsunami deposits generated by a meteorite impact into the late Eocene Atlantic Ocean 200 miles away near the head of Toms Canyon, 90 miles east of Atlantic City, N.J. Subsequent borehole evidence indicates that the tsunami deposits stretch from North Carolina to New Jersey.

most of the rock debris into a thick deposit (the Exmore beds) on the shallow seafloor (fig. 24).

Considering the evidence we had compiled, this was a plausible, yet audacious hypothesis. Should I publish it? I was a micropaleontologist/marine geologist, with no meteoritical pedigree, yet I was proposing that a thousand-foot-high wall of water was generated by a 35-million-year-old meteorite strike off the coast of ancient New Jersey. Some of my USGS colleagues already thought my proposal was ludicrous.

But, right or wrong, the hypothesis would be testable, an important aspect of any credible scientific idea. It might spur further research and understanding of the Exmore breccia and similar deposits. Misinterpretations and rejected hypotheses routinely contribute to the advancement of scientific understanding. Still, I decided to be cautious. Before sticking my neck out too far, I followed a time-honored tradition by first presenting the interpretation orally. I used it as the theme of a lecture given to colleagues attending the annual meeting of the Geological Society of America, in San Diego, California, on October 24, 1991. Concurrently, the USGS issued a press release to reach a larger audience.

The tale of a gigantic prehistoric wave devastating the East Coast quickly caught the imagination of the news media, and word of the hypothesis spread widely. Its favorable reception boosted my desire to formally publish the hypothesis in a scientific journal. The reaction of my scientific colleagues was mostly positive, as well. So I wrote a short manuscript to describe the main aspects of the evidence and to offer preliminary interpretations. The article was accepted by the journal *Geology*, and published in September 1992, under the title, "Deep Sea Drilling Project Site 612 Bolide Event: New Evidence of a Late Eocene Impact-Wave Deposit and a Possible Impact Site, U.S. East Coast."

It was a reasonable hypothesis, given the limits of the data available to us, but important issues remained to be addressed. The great thickness of the Exmore beds, the lack of tektites in them, and the absence of equivalent tsunami beds closer to Toms Canyon (on the coastal plain of New Jersey, for example) still required explanation. I should have seen these issues as clues that I still didn't have quite the right impact scenario.

CHAPTER SIX

Chesapeake Crater Revealed

IN ORDER to accurately map the distribution of the Exmore beds between the four Virginia core sites, we needed some reliable seismic profiles that would show the subsurface structure beneath Chesapeake Bay. At this time, Debbie Hutchinson had not yet rediscovered Rusty Tirey's 1982 seismic profiles, so once again, as in the Toms Canyon investigation, I turned to the oil industry for seismic data. And once again, Texaco, Inc., turned out to be my principal benefactor. That company, in partnership with the Exxon Exploration Company, had been exploring for oil and gas in southeastern Virginia. Dave Powars learned that, as part of their exploration efforts, the companies had collected a grid of seismic reflection profiles across Chesapeake Bay. Several of the profiles ran between the four Virginia cores sites, and we expected them to tell us how thick the breccia was, how deeply it was buried beneath the bay, and whether or not it really was continuous between the core sites.

I was pleased to learn that Parish Erwin, who had been the primary Texaco contact for the Toms Canyon data, also had been a key player in Texaco's Chesapeake Bay investigation. After several months of negotiation, Parish sent us four seismic profiles from the bay. The first glimpse of those profiles showed much more than we had ever expected. There, buried beneath the lower bay, was a huge hole, a mile or more deep. Moreover, the hole had been excavated not only into unconsolidated sedimentary rocks, but also deeply into the hard granitic rocks of the basement (fig. 25). A wide, flat, annular trough encircled the basement excavation. A steep fault scarp formed the outer rim of the annular trough, and a low-relief peak ring formed its inner rim. In combination, these were the principal structural elements of a complex impact crater that stretched more than fifty miles from rim to rim (fig. 26).

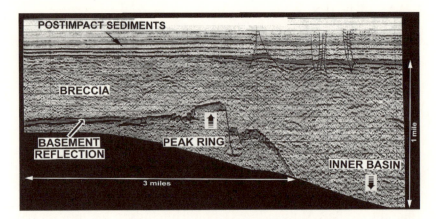

POSTIMPACT SEDIMENTS

BRECCIA

BASEMENT
REFLECTION

PEAK RING

INNER BASIN

3 miles

1 mile

25. One of the Texaco seismic reflection profiles, which clearly defines the southern margin of the peak ring of the Chesapeake Bay impact crater. The peak ring forms the steep wall of the deep inner basin. Chaotic reflections arising from large blocks and boulders in the Exmore breccia contrast starkly with the strong pair of basement reflections and the parallel, horizontal reflections coming from formation boundaries in the postimpact sediments.

At last, the ancient secret of Chesapeake Bay was revealed. Bob Mixon, Dave Powars, and Scott Bruce had unintentionally drilled into a crater three times larger than any other in the United States, the largest since the death of the dinosaurs, and the sixth largest on the planet. The main stratigraphic units displayed on the seismic reflection profiles matched precisely the vertical succession and thickness of individual sedimentary formations encountered in the core holes. On each profile, the Exmore beds displayed the chaotic reflection signature of impact breccia. Here was proof that the Exmore beds were not merely a super-tsunami deposit derived from the Toms Canyon impact. Instead, they were part of an immense breccia deposit generated directly by the impact of an enormous meteorite, which had struck right in the very center of our Virginia study area. Ground zero was almost precisely beneath the town of Cape Charles, between the Exmore and Kiptopeke core sites. By chance, two of the core holes had penetrated the breccia inside the crater, and two had sampled the ejecta blanket outside the crater.

It is humbling to contemplate the force required to excavate a 50-mile-wide crater to a depth of more than a mile, and then fill it back up again with breccia, all within a couple of minutes or hours,

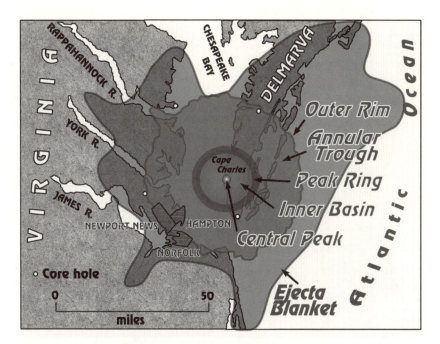

26. The Chesapeake Bay crater, centered over the town of Cape Charles on the Delmarva Peninsula, displays all the typical structural and depositional features of a complex peak-ring impact structure.

or at most, a few days. Computer simulations and comparisons with nuclear explosions show the rapidity of successive events in impacts of this magnitude. By similar analogy, we could estimate that the meteorite would have been 2–3 miles in diameter. It approached Earth at a velocity of around 60,000 miles per hour. After two to three seconds' passage through the atmosphere, it punched into the seafloor with a force equal to 10 trillion tons of TNT—a natural holocaust of colossal proportions. The immediate results included a crushing, supersonic shock wave rumbling through the Earth's crust, a super-heated blast wave tearing through the atmosphere, a gigantic surge of ejected debris roaring across the seafloor, a high-velocity curtain of crushed, melted, and vaporized rock debris spewing radially into the atmosphere, and a series of towering tsunami waves attacking the adjacent coastal plain. The blast wave alone would have instantly incinerated all higher life forms within six hundred miles of the impact site.

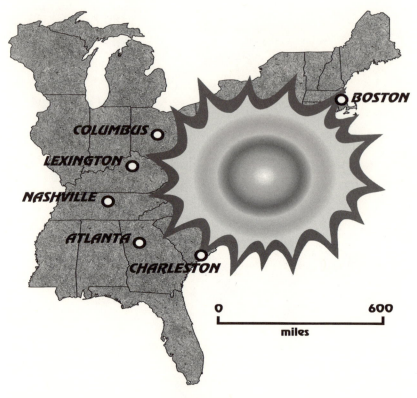

27. A 600-mile-wide zone of complete devastation would result if a 2- to 3-mile-wide meteorite were to strike Chesapeake Bay today.

A similar strike in Chesapeake Bay today would wipe out all the major East Coast cities, killing tens of millions. It would destroy the main centers of government and commerce, and severely cripple the U.S. economy (fig. 27). The enormous volume of dust ejected into the atmosphere would spread over the globe as a light-filtering screen, obscuring the sun and lowering temperatures at the surface for weeks to months. The scale of annihilation of wildlife, domestic animal herds, crops, and civilization as we know it, is appalling to contemplate.

I was thrilled to have the new insight provided by the seismic reflection profiles, even though it meant that I must revise my original interpretation. After all, the crux of the scientific method is to routinely replace old hypotheses with more plausible ones, whenever additional data demand it.

CHAPTER SEVEN

Proof in the Pudding

THE NEW seismic profiles allowed me to revise the original tsunami hypothesis, and to reinterpret all the geological anomalies Cede Cederstrom had discovered in southeastern Virginia. In collaboration with coauthors Dave Powars, Larry Poppe, and Bob Mixon, I published the revision as another article in *Geology*. In the August 1994 issue, under the title, "Meteoroid Mayhem in Ole Virginny: New Evidence of an Impact Crater Beneath Chesapeake Bay and Possible Source of the North American Tektite Strewn Field," we announced the discovery of a giant meteorite crater beneath Chesapeake Bay.

The revised interpretation garnered even greater excitement from the media than the original tsunami wave scenario. In the London *Sunday Telegraph* of September 4, 1994, Adrian Berry entitled his article "Smash hits from space." It was accompanied by an illustration showing King John and his knights thundering along on horseback, as an incoming asteroid chased them through breaking surf along a rocky beach.

The new interpretation clearly was a more plausible hypothesis than the first one. Nevertheless, it evoked a healthy dose of skepticism from both the press and the general scientific community. A front-page article in the *Baltimore Sun* proclaimed, "Giant crater may hold secret to bay's birth." However, Doug Birch, the *Sun's* science writer interviewed one prominent crater expert who was not impressed by the seismic profiles we published. That critic opined (and I agree) that interpretations of seismic profiles are often too subjective to be accepted at face value. As with abstract paintings, different viewers frequently see different things in the wiggle traces of the profiles. Furthermore, the sparsity of shocked quartz left a reasonable doubt in his mind that the Chesapeake

Bay crater was truly an impact structure. He was firmly wedded to a model that demanded an abundance of shock alteration. He offered no alternative explanation, however, for the Chesapeake Bay structure.

Another critic cited the lack of a central uplift as a major weakness in the impact interpretation. Nonetheless, he claimed to be 90 percent certain that I was right. "If it's not an impact crater, what is it?" he asked. In the journal *Science,* Richard Kerr's article carried the headline "Making an impact under the Chesapeake." Kerr interviewed a coastal-plain expert who claimed that the ". . . first reaction when you hear about something like this is to doubt it." Another crater specialist insisted that if I wanted to prove that the feature is an impact crater, I would have to drill in the center of it.

Nevertheless, this was a time of unparalleled excitement for me and my collaborators, within and outside the USGS. Up to this point I had spent thirty-five years publishing more than two hundred articles and books. But not one of them had provoked as much public enthusiasm and discussion as this single article about a mysterious cosmic convulsion that took place deep in the planet's murky past.

A television program that I particularly enjoyed contributing to was funded and produced by the Prince William County Public School System in Virginia. Jon Bachman, a teacher and curriculum specialist for that system, organized and narrated an hour-long "satellite field trip," devoted to the geology and ecology of Chesapeake Bay (fig. 28). The purpose was to take viewers on a field trip via a real-time satellite uplink from video cameras on the ground. In this case, the show was telecast live from the stern of a fishing boat on the Potomac River, just upstream from its junction with Chesapeake Bay, not far from the crater rim. More than 500,000 school children received the telecast directly in their classrooms.

That field trip provided a rare opportunity to speak directly to a host of young minds, eager to learn how science contributes to understanding our natural environment. I could demonstrate to them how different geological science is from worn stereotypes depicting wild-haired nerds in thick glasses and white lab coats.

28. A televised "satellite field trip" sponsored by the Prince William County School System of Virginia gave me an opportunity to present the Chesapeake Bay meteorite impact story directly to half a million school students.

Geological research gets you outside, tracing ancient rock forma-tions along mountainsides or riverbanks, digging for fossils in some remote wilderness—even chasing asteroids and comets. Ge-ology also gets you inside, drilling holes deep into the Earth and under the sea, surveying Earth's interior from computer-guided research vessels, or scanning the innermost structures of minerals and fossils with powerful electron microscopes. It also provides many opportunities to interact with colleagues all over the world. Geology is a global science. Geologists go wherever the rocks are.

During this period, I was especially gratified that so many peo-ple appreciated the results of research that was, in many respects, outside my field of specialization. Still, I was a newcomer to the disciplines of planetary geology and impact cratering. Some scien-tists resisted my initial interpretations partly because I had no prior credentials as a crater specialist—they didn't know who I was. Their acceptance of my hypothesis would (and should) have to depend on the merits of the science alone. I had to present an overwhelming case based on abundant diagnostic evidence. It helps sometimes, however, if some of the leading experts are attracted by the eloquence of one's ideas.

The first step toward solidifying the case was to acquire additional evidence of shock metamorphism among mineral constituents in the Exmore breccia. For this acquisition, I was fortunate to collaborate with Christian Koeberl, an internationally recognized expert on impact shock, and Professor in the Institute of Geochemistry at the University of Vienna, Austria. Chris contacted me to request samples of the breccia, so that he could look for evidence of shock. He also wanted to compare the chemical compositions of breccia components with constituents of late Eocene impact ejecta at other localities within the North American tektite strewn field. On the basis of his extensive experience with other impact deposits, Chris wanted to concentrate mainly on clasts of crystalline basement rocks. This approach contrasted with the initial studies by Larry Poppe and Glen Izett, who searched for individual grains of shocked quartz in the sandy matrix. Chris knew firsthand, that if we were truly dealing with a giant impact crater, most basement clasts would have recorded their shock history. We expected that if he could demonstrate this record at Chesapeake Bay, it would convince most remaining critics that the impact hypothesis was sound.

So I selected fragments of granitic and metamorphic basement rocks for Chris (fig. 29). First I had to boil the samples and wash all the clay out of them using a fine-mesh screen. I then dried the samples and sifted each through another set of wire-mesh screens to separate all particles larger than one-sixteenth of an inch. Next, using forceps and a low-powered optical stereomicroscope, I extracted all basement fragments from each subsample. Most were smaller than 0.04 inch. I was surprised to find that the tiny fragments were rather common in the screened residues, whereas they had not been obvious in hand-sized sections of core. I sent Chris basement fragments from sixty-five different levels in the Exmore breccia.

When the samples arrived in Vienna, the individual particles were still too large for Chris to analyze properly. First he had to

Basement clasts

Windmill Pt.
516.00 —.25'
FLOAT

cwp

Foraminifera

29. I used forceps and a tiny paint brush to extract quartz grains, minute clasts of crystalline basement, and fossil foraminifera from a sieved sample of Exmore breccia. Chris Koeberl analyzed the grains and clasts under a high-powered microscope to identify shock deformation features.

slice them into wafers only a few thousandths of an inch thick, which he glued to a thin glass microscope slide. By passing light through the thin sections on the stage of a high-powered optical microscope, he could determine their mineralogical composition and document shock features. Chris enlisted two additional collaborators for this study: Uwe Reimold and Dion Brandt, geologists with the University of the Witwatersrand, in Johannesburg, South Africa. Just as Chris had expected, the basement clasts were loaded with shocked grains. Most shocked grains were quartz and *feldspar*, two common constituents of granitic rocks. Other grains showed signs of melting around their edges, another diagnostic result of the tremendous heat built up by impact shock. These results gave us indisputable new evidence that the Chesapeake Bay structure was a giant impact crater.

While Chris, Uwe, and Dion scurried about their labs in Vienna and Johannesburg, I decided to request more seismic reflection profiles from Texaco. Their initial release of data to the USGS had included only four of many profiles they had collected from the bay. None of those original four had come close to the center of the crater, which underlies the town of Cape Charles, on the western shore of the Delmarva Peninsula. So I asked for additional profiles that crossed the northern rim of the crater and approached nearer to Cape Charles. The Texaco geologists were intrigued by my interpretations, having been confounded by the mysterious "hole in the basement," when they had first analyzed the seismic profiles. Because their interpretations had been strongly influenced by conventional concepts of coastal plain evolution, they, like previously cited researchers, had not considered an impact origin for the feature. But they fully supported my concepts, and Parish Erwin sent five additional profiles.

As soon as I examined the new seismic profiles, I knew that, in combination with the impact-shock data from Chris, Uwe, and Dion, we had enough diagnostic evidence to convince the most reluctant skeptics that the Chesapeake Bay crater was indeed America's largest meteorite impact crater. Two of the new seismic profiles displayed the basement excavation so graphically that even a novice seismic interpreter could not miss it (fig. 30). Furthermore, the new profiles showed convincingly that a distinct, raised, peak ring encircled the deep central basin. In composite, the peak ring was an irregular, ovate ridge of low peaks, about six hundred feet high at maximum relief. With all these seismic data, we could construct a computer-generated, three-dimensional model of the crater (fig. 31). The model displayed all the diagnostic structural features more realistically than the two-dimensional maps.

Supported by this new diagnostic evidence, Chris and I were eager to fortify my hypothesis and win over the remaining skeptics in the meteoritical community. First, we would submit a coau-

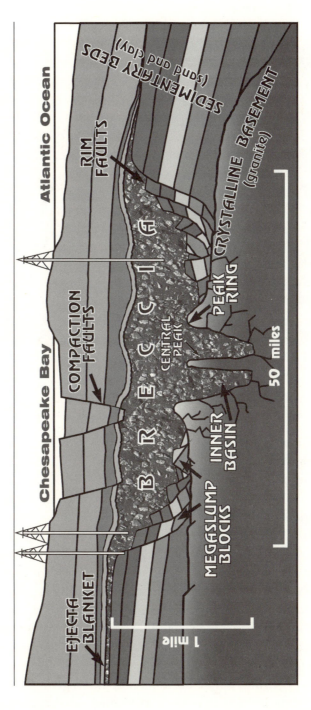

30. Interpretive cross section of the Chesapeake Bay impact crater based on seismic reflection profiles and core holes drilled in southeastern Virginia. All the structural and depositional features of a complex peak-ring crater are clearly documented.

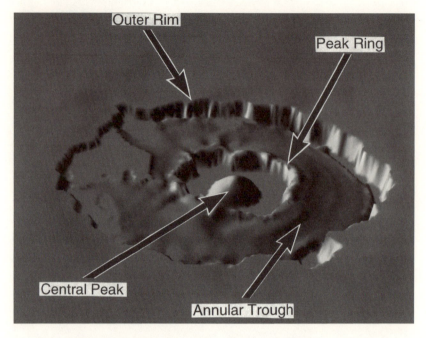

31. Three-dimensional perspective of the Chesapeake Bay crater with the breccia removed. This image was produced by a computer using data derived from seismic reflection profiles. Compare with King crater in figure 19. Image courtesy of Michael Taylor.

thored article to *Science*, America's premier weekly research magazine aimed at the general scientific community. This article would emphasize Chris's verification of impact-shock evidence in the Exmore breccia, so he would be principal author. I would submit a separate single-authored article to *Meteoritics and Planetary Science*, a leading journal for crater specialists. In this paper, I would discuss the implications of the new seismic profiles. Meanwhile, as these articles were being reviewed and revised for publication, we had an opportunity to present our new findings in person, at the annual meeting of the Meteoritical Society. This society's membership includes the inner circle of meteorite specialists. As luck would have it, the Society was meeting at the Smithsonian Institution in Washington, D.C., a perfect place to confront doubters and to rekindle public excitement about a giant meteorite crater nearby under Chesapeake Bay. Science writers from both *Science* (headquartered in D.C.) and the *Washington Post* were sure to attend the meeting.

On Monday, September 11, 1995, we presented our case. The meeting took place in a beautiful underground conference facility constructed deep beneath Washington's Capitol Mall, the broad park that runs between the Capitol building and the Lincoln Memorial. As a result of our presentations, we felt confident that none of the attending delegates had further reservations about the crater's origin. The Chesapeake Bay crater was solidly documented with indisputable diagnostic evidence.

Immediately following our oral presentations, we were invited to discuss our ideas with Justin Gillis, a staff writer for the *Washington Post*. The following day, Justin's article made front-page headlines, asserting that "Bay Meteor Theory Appears Rock Solid" (fig. 32). Poag and Koeberl had offered new evidence of an ancient cataclysm, which clinched the case for the Chesapeake Bay crater. The Associated Press distributed Gillis's article worldwide. Even my hometown newspaper, the *Cape Cod Times*, carried a condensed version. Ironically, theirs was the only article about the crater that didn't credit the discoverers by name. Ten days later, in the September 22 issue of *Science*, Richard Kerr also proclaimed confirmation that the Chesapeake Bay crater had been generated by a giant meteorite impact.

Richard Grieve, a geophysicist and crater specialist with the Geological Survey of Canada, and keeper of the semiofficial list of Earth's impact craters (also an early critic of my hypothesis), was convinced enough to add Chesapeake Bay to his roll call. Richard also was an assistant editor for *Meteoritics and Planetary Science*. In that capacity, he graciously explained some of the subtle implications of the seismic interpretations, and helped immensely in revising my manuscript.

At last the meteoritical community was satisfied that a giant meteorite had indeed struck the U.S. East Coast in the late Eocene. The physical results of the bombardment were unequivocally recorded in the cores and seismic profiles. The size, shape, and principal structural elements were clearly identified and defined. It was time now to consider the geological and paleoenvironmental

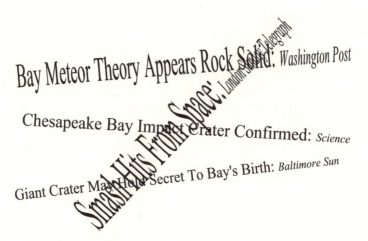

Bay Meteor Theory Appears Rock Solid: *Washington Post*

Smash Hits From Space: *London Telegraph*

Chesapeake Bay Impact Crater Confirmed: *Science*

Giant Crater May Hold Secret To Bay's Birth: *Baltimore Sun*

32. Collected headlines from around the world, following exposure of our hypothesis to the media.

consequences and implications of this violent collision. What happened to life forms in the impact area? Is there evidence of a massive late-Eocene die-off in eastern North America? Did the impact cause a mass extinction like that of the dinosaurs, 30 million years earlier? We can start to answer these questions by investigating what is known about the geological and paleontological record of the Eocene Epoch. Were the continents and oceans similar to today? What was the climate like? What terrestrial and marine organisms populated those ancient environments?

CHAPTER EIGHT

A Perilous Path

THE LATE Eocene Earth, 35 million years ago, was home to a diverse assortment of familiar and unfamiliar organisms. Odd ancestral mammals dominated heavily forested terrestrial habitats; birds and insects flourished in warm, moist, tropical skies. Profuse populations of bony fish filled the oceans, lakes, and rivers, while mussels, clams, oysters, starfish, and corals filled the bottom habitats. However, these complex organisms were actually new kids on the block. They were newcomers, because it took evolution nearly 4 billion years to turn primitive aquatic bacteria into the complex, interactive communities of the Eocene (fig. 33). Thirty-five million years ago may seem like an eternity to us short-lived humans, but it was only yesterday in the grand scheme of organic evolution, only 0.9 percent of life's existence on Earth. If, for the purpose of analogy, we imagine that life on Earth began only one year ago, the late Eocene would have begun only three days ago. (On this timescale, modern man's presence on the planet would be a mere four hours!) To emphasize this point, let me review briefly the remarkable journey that brought such bountiful life to the Eocene. My review is, of necessity, abbreviated; I mention only a few highlights. Relating nearly 5 billion years of Earth history in a few pages of text is akin to writing my autobiography using only three words. If you would like a more detailed and balanced summary of evolution's perilous path, please turn to Richard Fortey's and Stephen Jay Gould's exciting accounts, listed as recommended reading at the end of this book. As you read life's story, however, it is important to remember this. Our knowledge of geologic history and organic evolution is constantly changing as new data are collected and new discoveries are made. The broader concepts are less subject to revision than the finer details, but surprises keep turning up. Of special importance in this regard, China has opened up vast new areas to western geologists and paleontologists. Results from the past ten years of fieldwork there have already changed many long-held ideas ranging from the cellular development of Proterozoic organisms to the evolution of birds from dinosaurs. There is much more to come.

The best evidence from rocks and fossils indicates that Earth formed about 4.7 billion years ago. For the first 700 million years, our planet was a barren wasteland, totally devoid of life. It was a desolate, volcanic globe of hot rock and water, enveloped by a gaseous atmosphere. The decay of radioactive elements produced interior heat, which was probably five times greater than it is today. The ocean, too, was warm, a chemical stew in which inorganic chemical reactions were the principal means of energy production and exchange. These early lifeless years have been called the Age of Chemistry.

Life's improbable evolutionary journey started around 4 billion years ago, when organic molecules in the primitive ocean arranged themselves into the first bacteria. Each individual bacterium was a microscopic bit of jelly-like cytoplasm packaged inside a single cell, which lacked the chromosomes and nucleus characteristic of higher organisms. Mammal cells, for example, contain as many as six hundred chromosomes; they reside within a distinct spherical nucleus, which is separated from the cell's other contents by a gelatinous membrane.

The global environment at this stage of Earth history was hostile to most life as we know it, especially because the atmosphere contained no free oxygen. Instead, it was a noxious mixture of mainly ammonia, carbon dioxide, methane, and hydrogen sulfide. When filtered through this miasma, the light of the early Sun would have turned the sky pink. But of course, no eyes were there to see it.

The chemistry of the primordial ocean was also much different than today. Its makeup would have depended on the types of minerals eroding from the continents and the atmospheric gases that dissolved in it. Lacking a source of free oxygen, the ocean would have been oxygen deficient, and probably was more acidic than at present.

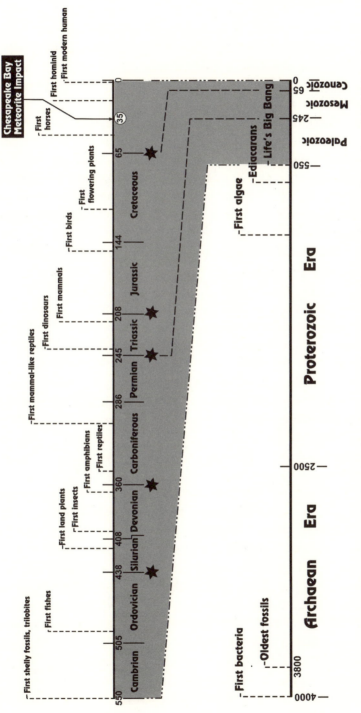

33. Scaled diagram (millions of years) showing the relative amount of time represented by geologic eras and periods, and highlighting important steps in life's perilous 4-billion-year journey. Stars indicate positions of the Big Five mass extinctions.

The archaic bacteria were tough enough to endure these harsh conditions, because they needed only inorganic substances, such as hydrogen and sulfur, for their life processes. Gradually though, a few ancestral bacteria evolved into somewhat more complicated multicelled types. Some of them, the *cyanobacteria*, acquired the ability to convert the sun's energy into a nutrient source—they invented photosynthesis. Luckily for us, the principal waste product from photosynthesis is oxygen. Eventually, oxygen produced by bacterial photosynthesis fundamentally altered the chemical makeup of the atmosphere and changed the subsequent course of life on our planet. But this atmospheric conversion didn't happen overnight—it took 3.5 billion years, or 75 percent of Earth's entire history. You shouldn't be surprised that it took so long. Just imagine the infinitesimally small volume of oxygen that an individual bacterium or even trillions of bacteria can produce.

Once free oxygen (O_2) was abundant, ozone (O_3) also increased. You may be familiar with the importance of ozone in the atmosphere. The news media have widely publicized the ozone "hole" that has developed over Antarctica in the last decade. The absence or thinning of the polar ozone layer creates a portal through which cancer-causing ultraviolet radiation can reach the surface virtually unimpeded. An increase in ozone volume did just the opposite for primordial Earth. It helped to screen out lethal ultraviolet radiation. This allowed microbes, which formerly were limited to aquatic environments, to colonize the barren land. So bacteria flourished in these ancient poisonous or almost sterile environments for 3.5 billion years, while they evolved imperceptibly into more complex types. Cyanobacteria were so abundant and dominant from 2.5 billion to 550 million years ago, that this interval is often called the Age of Bacteria. We have many good fossil examples of these primitive cells. More than 3,000 different observations of individual microbes have been recorded from rocks of the earliest two geological Eras, the Archaean (3.8–2.5 billion years ago) and the Proterozoic (2.5 billion to 550 million years ago).

To me, the most fascinating fossil bacteria are not the individual specimens, but massive accumulations of them, which built stony structures called *stromatolites*. Stromatolites formed in shallow coves and bays where the bacterial soup coagulated into thin, slimy films. The surfaces of these films were sticky enough to collect thin layers of dust and sedimentary debris. Each paired layer of bacteria and debris constituted a *microbial mat* less than 0.04 inch thick. Because there were no predators or other higher organisms around to burrow into or chew up these mats, they could last for immense periods of time. In some cases, hundreds of additional microbial mats accumulated on top of the original mat like a stack of paper-thin pancakes. Through time, the stacked mats solidified into thick clumps, mounds, or columns. You can slice right through a stromatolite mound with a rock saw and count the layers. The largest stromatolites discovered so far contain thousands of layers, and are three hundred feet thick.

This primitive type of bacterial structure is truly a triumph of biotic engineering. It has withstood every form of calamity and environmental stress the Earth has suffered for three billion years, yet modern stromatolites remain virtually unchanged from their ancestral state. Paleontologists understand the development of stromatolites very well, because living examples, 1–2 feet high, inhabit shallow marine waters in a few isolated locations, like Shark Bay, Australia; Baja California; and the Bahamas.

Though bacteria reigned supreme throughout most of the Earth's early history, they were not always entirely alone. Sometime in the late Archaean Era, the initial forms of life with nucleated cells evolved into simple microscopic plants—marine algae. Fossils of diverse species of marine algae are well known from Proterozoic rocks as old as one billion years.

One of the biggest puzzles of the late Proterozoic fossil record is the true identity of a group of jellyfish-like body impressions, known as Ediacarans. The name is taken from their place of discovery, the Ediacaran Hills in South Australia. These strange organisms existed only for about 10–30 million years, just prior to the beginning of the Paleozoic Era. But they were part of an abundant, diverse, aquatic community, which was distributed globally. Ediacarans range in size from less than half an inch to about three feet. Their fossils are shaped like scalloped discs, spoked wheels, leaves, worms, and even Oreo cookies (fig. 34). But only impressions of their soft bodies, never any hard parts, have been found. No functional organs, like legs, eyes, mouths, or stomachs can be discerned, though body segmentation and exterior ornamentation is excellently preserved. So far, their relationship to any other extinct or modern animal group is unknown. We don't even know whether to classify them as animals or plants. One German researcher, Dolf Seilacher, has gone so far as to put them in a separate phylum or kingdom of organisms (neither animals nor plants), a sort of "failed experiment" of evolution, which died out leaving no descendants.

To a time-traveling visitor from our action-driven civilization, the Archaean Era and most of the Proterozoic would have been intolerably boring, regardless of the magnificent stromatolites. With inanimate bacteria and algae dominating the land and seascape for 3.5 billion years, the human eye could detect no living activity. Practically the only discernable action would have been winds, waves, and the eruptions of volcanoes and geysers.

The first real live action from crawling, swimming, oxygen-breathing, hard-skeletoned, multicellular animals didn't take place until life's journey was almost 90 percent complete. It happened suddenly, about 550 million years ago, at the beginning of the Cambrian Period, the oldest subdivision of the Paleozoic Era. But after such a long delay, that action was worth waiting for. In the space of a mere 15 million years (about 0.4 percent of the first 3.5 billion years), life exploded into a vast worldwide community of hard-bodied and soft-bodied marine invertebrates. These were real animals, though none had backbones. This initial community left an enormous collection of fossils in the geologic record. Life as we know it, like the universe itself, started with a "big bang." It all took place underwater, however. The Cambrian continents still were completely uncolonized by anything other than bacteria and algae.

Trilobites are the most famous and widespread hard-bodied invertebrate fossils of the early Paleozoic Era. Their three-lobed, multi-legged, arthropod bodies were protected by stiffened external skeletons as they competed for living space on the Cambrian

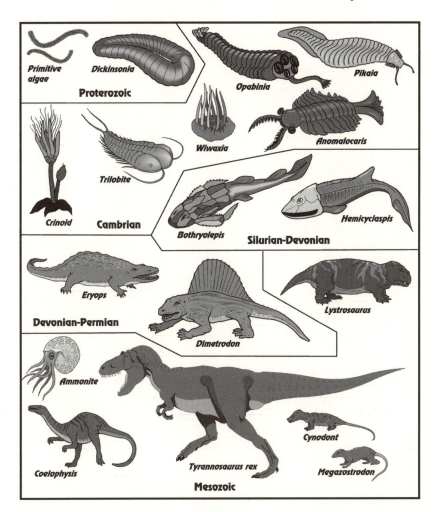

34. These reconstructed fossil organisms represent some of the key players and some of the dead ends in evolution's route to the Age of Mammals. *Dickinsonia* was an Ediacaran; *Opabinia, Pikaia, Wiwaxia,* and *Anomalocaris* were bizzare members of the Burgess Shale assemblage; *Bothryolepis* and *Hemicyclaspis* were primitive fish armored with thick bony plates; *Eryops* was a primitive amphibian; *Dimetrodon* was a primitive mammal-like reptile; *Lystrosaurus* was the principal land animal remaining after the Permian mass extinction; *Coelophysis* was a relatively small (9-ft-long), bipedal, carnivorous dinosaur from the Late Triassic Period. Cynodonts were advanced mammal-like reptiles; *Megazostrodon* was a shrew-like primitive mammal.

seafloor (fig. 34). Trilobites proliferated into about 10,000 species during the 400 million years of their existence.

Another successful Cambrian animal group were the sea lilies, or *crinoids*. Most of these tulip-shaped organisms were attached to the seafloor by a long, slender stalk (fig. 34). The stalk was flexible, essentially a succession of small limey disks stacked up like checkers. These disks separated and scattered when a crinoid died, and Paleozoic limestones are full of them. At the top of the crinoid stalk was a blossom-like head, or *calyx*. Numerous waving tentacles sprouted from the calyx and helped the crinoid to catch food particles from the water. More than 5,000 extinct species of crinoids have been discovered, some of which grew to the enormous length of 60 feet. Over 800 species still live in today's oceans. Most modern forms are *feather stars*, which have lost their stalks; they can swim about, independent of the seafloor.

Trilobites, most crinoids, and many other groups of Cambrian animals are exotic and extinct, but the Cambrian seafloor habitats also supported ancestral representatives of familiar animal groups. Corals, sponges, snails, and clams were well represented, and primitive relatives of sea urchins, squids, and fish evolved during this interval, as well.

Cambrian seas also contained a profusion of soft-bodied animals, species without any hard, mineralized structures. The squashed remains of these creatures have been preserved in fine-grained sedimentary rocks as thin organic films and impressions. With meticulous care, paleontologists can take these films apart, layer by layer, and reconstruct the animals. Many of these squishy Cambrian organisms had odd arrangements of legs, mouths, and other body parts, unlike anything we know today. They looked like rejects from Dr. Frankenstein's laboratory. One of the most famous places to collect soft-bodied and delicately skeletonized Cambrian fossils is an outcrop of dark, clay-rich rock, called the Burgess Shale. The Burgess Shale occurs in a small quarry, eight thousand feet up in the Canadian Rockies in eastern British Columbia. A variety of small ancestral arthropods, worms, sponges, clams, and sea urchins have been found there. Besides the familiar groups, however, the Burgess Shale has yielded twenty bizarre species that defy classification. They don't fit into any established phylum. Even their names, many derived from Indian words, are especially exotic—*Wiwaxia, Takakkawia, Yohoia, Opabinia,* and *Anomalocaris* are among my favorites (fig. 34). These animals are good examples of evolutionary dead ends. We don't know their predecessors, and they left no recognizable descendants.

The remaining 305 million years of Paleozoic time introduced most of the main groups of marine and terrestrial plants and animals familiar to us, but each of its five subdivisions (periods) had its own distinctive flora and fauna. For example, the Ordovician Period, which followed the Cambrian, is notable for the tremendous abundance and variety of *brachiopods.* These two-shelled, bottom-dwelling animals were the early Paleozoic competitors of clams. In fact, they looked like clams, except that most were attached to the bottom, similar to mussels. Brachiopods initially outnumbered clams, and radiated into more than 30,000 species. They eventually lost out to the clams, although more than 260 species of brachiopods still exist. Living brachiopods can be found in all oceans, including polar seas.

Invertebrate marine life remained much the same during the next two Paleozoic periods, the Silurian and Devonian. One notable change among the higher aquatic animals, however, was a spurt of evolution among marine and freshwater fish. Devonian rocks are particularly rich in fish fossils (fig. 34). Fish were so abundant and diverse during the remainder of the Paleozoic that paleontologists often call this the Age of Fishes. Earliest fishes lacked movable jaws, which limited their diets to things they could filter out of the water or suck up off the bottom. Their bodies were heavily armored with large bony plates rather than thin scales. Modern-looking, jaw-bearing fish, with either cartilaginous (shark-like) or bony internal skeletons, were scarce until the late Devonian and Carboniferous Periods.

Perhaps the most significant evolutionary event of the middle Paleozoic was a full-scale invasion of the land by both plants and animals. Once the invasion started, it didn't take long for the continents to acquire a thick cover of plants. Dense forests of tall trees, called *lycopods,* and 50-foot tree ferns spread widely, and more than

10,000 species of other primitive leafy plants flourished in thick jungles and swamps. The lycopods were 100–150 feet high, with bare trunks and bushy crowns of leaf- and spore-bearing branches. The swampy late Paleozoic environments preserved enormous volumes of dead plant debris, which were eventually transformed into thick deposits of coal. Rocks of the Carboniferous (carbon-rich) Period provided the coal that fueled the industrial revolution.

These lush continental habitats were soon invaded by platoons of air-breathing arthropods—insects, spiders, and scorpions. The earliest insects were mainly small crawlers, like centipedes, millipedes, and mites. Some of the primitive insects, however, quickly found an ingenious way to escape ground-crawling predators. They took to the air on the world's first-known wings. Flying birds and bats were still tens of millions of years in the future, but there were plenty of hungry amphibians, reptiles, spiders, and competing insects who liked to crunch up juicy insect bodies.

Carboniferous insects gloried in their diversifying lifestyles and some reached gigantic body proportions. Six-foot millipedes and four-inch cockroaches burrowed and crawled in the organic debris littering the great swamps and forests. Above them, dragonflies with two-foot wingspans were the "Red Barons" of the late Paleozoic airways.

The first of our distant, amphibious, four-legged or *tetrapod* ancestors, also crawled out of their aquatic habitats during the Devonian Period. As soon as archaic amphibians mastered the skills of walking and breathing oxygen, the race was on for dominion over the land. Amphibians enjoyed a short-lived dominance during the Carboniferous, but in the early part of that period, the first primitive reptiles arose, poised to eventually command all terrestrial habitats for 150 million years. Among the early reptiles were forms with mammal-like features, such as high skulls, powerful jaws, and long, slashing teeth. Some of these mammal-like reptiles, such as the well-known *Dimetrodon*, sprouted high, spiny "sails" down the middle of their backs, to help regulate their body temperature (fig. 34). Descendants of these mammal-like reptiles led to the evolution of humans—*Homo sapiens.*

Early reptiles devised a new, more efficient way to protect their offspring—the amniotic egg. Amniotic eggs were the first to have a tough, leathery or mineralized shell. This protective sheath isolated reptile embryos from outside disturbances until they reached full development. Furthermore, these eggs could be laid anywhere a parent chose. The increased efficiency of this incubation system gave reptiles a distinct advantage over amphibians, who had to return to the water to lay their soft, gelatinous egg masses.

Reptiles proliferated in the late Paleozoic Era (Permian Period) and the early part of the Mesozoic Era (Triassic Period). The most famous of all reptile groups, the dinosaurs, appeared in the early Triassic as small creatures that walked upright on their hind legs (fig. 34). But it took another several million years for the dinosaurian giants to arise. The first of these imposing beasts appeared in the early part of the Jurassic Period, roughly 200 million years ago.

The Jurassic is probably the best publicized of all geological periods, because of the huge success of Michael Crichton's novel, *Jurassic Park*, and its blockbuster movie version. Beyond its sci-fi fantasies, Crichton took great liberties with his dinosaur tale, because many of the principal dinosaurs in his narrative, including the awesome *T. rex*, *Velociraptor*, and *Triceratops*, did not live in the Jurassic Period. Instead, they were prominent members of the late Cretaceous dinosaur community, which closed out the Mesozoic Era, 65 million years ago.

While the reptiles had their way on land, the dominant sea creatures were an exotic group of highly successful squid-like cephalopods, called *ammonites* (fig. 34). Ammonites were the most abundant and widely spread marine invertebrates of the Mesozoic Era. Their shells were rolled up into a flat coil, so that they resembled automobile tires. The animal added larger and larger chambers as its body grew. The ammonite lived in the final chamber of the shell, and looked very much like the modern *Nautilus*, with large humanoid eyes, beak-like jaws, and flexible tentacles. The largest ammonites were gigantic beasts, whose shells were nearly seven feet in diameter.

The Mesozoic Era also is notable for another milestone along the evolutionary pathway of terrestrial plants. In the early part of the Cretaceous Period, the first *angiosperms*, or flowering plants, evolved. They quickly spread out into almost all terrestrial environments, where many species developed a dependence on bees, birds, and eventually bats, to pollinate their flowers. Thereafter, the evolution of angiosperms and their pollinators proceeded hand in hand. Characteristics such as flower shapes and plant reproductive cycles changed in unison with the evolution of specialized mouthparts and reproductive strategies of the insects.

Though the end of the Cretaceous Period brought the reign of reptiles to a close, a few species, like crocodiles, turtles, snakes, and lizards, crossed into the Cenozoic Era. In contrast, none of the ammonites made it across. Some of their distant cephalopod cousins did, however. They evolved into the squids, octopuses, and nautiluses we know today, but they never regained the prominence of their ammonite predecessors.

The dawning of this youngest geologic era opened the evolutionary gates on land for a group of small, furry, milk-producing, warm-blooded, ancestral mammals (fig. 34). Beginning with nocturnal shrew-like creatures, these invaders quickly filled most ecological niches vacated by the reptiles, and established dominion in the Age of Mammals. At least we chauvinistic mammals like to think this is our age. Some scientists argue that it is more properly called the Age of Insects, for these "lowly" organisms far outnumber mammals. But if numbers count, we might as well call the whole of life's existence the Age of Bacteria and be done with it. Regardless of our choice of terminology, entry into the Cenozoic Era brings us back to the Eocene and the setting for the Chesapeake Bay impact.

The evolution of life through billions of years of geologic time was brilliantly formulated by Charles Darwin in 1859, and fervently promoted by his friend and colleague Thomas Henry Huxley in the middle and late 1800s. From 1859 onward, the evolutionary process was generally envisioned as a gradual change in bodily features and functions through geologic time. This slow pace allowed complex new species to develop from old primitive ones in a procession of biologically related lineages. As paleontologists filled in many of the gaps in the fossil record and the precision of rock-dating methods improved, they realized that evolution operated on a wide variety of timescales. Many lineages changed gradually, but others remained relatively unchanged, or in equilibrium, for millions of years. Then, suddenly, that equilibrium would be punctuated by a significant change, often a dramatic one. They learned that most of life's lineages had to contend with a never-ending succession of filters, bottlenecks, blind alleys, and extreme calamities brought about by changes in their environment. Traditionally, the severest environmental stresses were thought to have been caused by episodic shifting of continental plates, cycles of rising and falling sea level, the buildup and melting of glaciers and ice sheets in polar regions, and intervals of exceptionally intense volcanic activity. It was not always obvious which one or combination of these prime forces caused a particular environmental shift. It was quite clear, however, that the most rapid shifts happened too quickly for many of the organisms to adapt to the new environments. This jolted the underpinnings of plant and animal communities alike, and in a few instances, nearly eliminated all higher life forms. We call these jolts *mass extinctions.*

CHAPTER NINE

Mass Extinctions

Up until a few years ago, it was considered outrageous to suggest that any of life's mass extinctions could be connected in any way to a meteorite impact. Though the idea was first proposed in the mid-1700s, by French scientist Pierre de Maupertuis, a lack of convincing field data left most geologists and paleontologists highly skeptical. But this attitude has largely disappeared because of the innovative work of two brilliant researchers—Gene Shoemaker and Walter Alvarez. Gene and Walter worked for different organizations, and took different approaches to meteorite research. Together, however, they instigated a revolution in our understanding of meteorite impacts and their consequences. These two men and their colleagues have shown us that impacts are fundamental processes in the development of all planetary bodies in our solar system. Furthermore, Gene and Walter have demonstrated how extraterrestrial impacts modify the pace of Earth's organic evolution through the mechanisms of global change and mass extinction.

Gene Shoemaker's contributions to impact studies are too numerous to mention in a few paragraphs. He essentially invented this field of research. Working as a geologist for the USGS, Gene established its Branch of Astrogeology in Flagstaff, Arizona, and served as its director for five years. Upon retirement from the USGS, Gene went to Caltech as Professor of Geology and Chairman of the Department of Geological and Planetary Sciences. During his distinguished career, Gene was showered with awards, among which was election to the National Academy of Sciences. Gene developed the key geological, geophysical, and geochemical methods used to prove that many of the large bowl-shaped geologic structures on Earth are genuine meteorite craters. Starting with a masterful analysis of Arizona's Meteor Crater (1956), he helped to discover the first natural occurrence of shock-generated coesite at that impact site. His field and laboratory methods, especially the use of reversed stratigraphy and shocked minerals, have been the keys to identifying more than 150 impact craters and numerous deposits of impact ejecta, both on land and on the seafloor. He also trained the lunar astronauts who brought impact debris back from the Moon. He was in NASA's Houston Control Center to personally direct lunar rock collections by the Apollo 11 and 12 astronauts.

In July of 1997, Gene's amazingly productive career was cut short by a fatal automobile accident. He was in Australia, examining some of the many impact craters that he had discovered in the vast outback. In a fitting tribute to his legacy, Gene's ashes were sent to the Moon in January 1998, aboard the spacecraft Lunar Prospector. After evaluating the mineral wealth of that airless globe for about a year from orbit, the spacecraft is programmed to crash into the Moon's surface, where Gene's remains will rest among a whole "planetful" of impact craters.

Walter Alvarez is Professor of Geology and Geophysics at the University of California, Berkeley, and, like Gene Shoemaker, a member of the National Academy of Sciences. Building on Gene's discoveries, Walter and a team of Berkeley colleagues proposed the first plausible, testable explanation for the sudden, worldwide extinctions at the K-T boundary. Ironically, Walter's breakthrough began with micropaleontology, as did mine, ten years later. Studies of abrupt foram extinctions at the K-T boundary in Italy and Denmark led him to find an unusual abundance of the rare element *iridium*. Curiously, a half-inch clay layer at both sites contained thirty to sixty times more iridium than rocks above and below the clay. Moreover, the iridium spike occurred precisely at the K-T boundary. Walter's father, Luis, a Nobel Laureate in physics, suggested that the iridium enrichment must signal an ancient asteroid impact. He knew that asteroids contain much higher concentrations of iridium than Earth's crust. Some quick calculations indicated that deposition of a worldwide layer containing this much iridium would require an asteroid of gigantic proportions— at least six miles in diameter. The resultant impact would have sprayed an enormous volume of dust particles into the atmosphere, blocking out sunlight and causing surface temperatures to plummet. A prolonged "impact winter" might account for the abrupt worldwide K-T extinctions. As soon as the Berkeley team (Luis, Walter, Frank Asaro, and Helen Michel) published their startling idea (1980), other researchers quickly began to report elevated iridium concentrations at K-T boundary sections all over the world.

A raging controversy over impacts and extinctions began to sweep like wildfire through the geological community. The fiery arguments abated somewhat when the Chicxulub crater was recognized as the K-T impact site, but embers of heated debate still smolder here and there.

In a nutshell, Gene Shoemaker, Walter Alvarez, and their followers demonstrated that large meteorite impacts are routine events in Earth's geologic history. Furthermore, they periodically alter global climate so severely and rapidly that the biosphere suffers catastrophic die-offs and mass extinctions. Though subdued debate about the details of extinction and impact processes continues, compelling new supportive evidence is piling up from nearly every new field study and computer model. Of course, many other geologists, paleontologists, and other scientists participated in the K-T research. They helped to elevate the impact-extinction hypothesis to a well-tested and widely accepted theory. In yet another good example of the way science works, the perceptive criticism of many disbelieving colleagues helped to weed out spurious data and interpretations. I urge you to meet some of these other researchers in the pages of Walter Alvarez's own book about this "crater of doom," and in Bill Glen's fascinating discussion of the extinction debates. I have listed their book titles at the end of this volume.

Paleontologists can count approximately ten to thirty mass extinctions in the fossil record, depending upon who is doing the counting. The postulated number and severity of extinctions depend in part on the level of classification, or taxonomic rank, at which the census is taken. Living organisms are ranked in a hierarchy of groups, or *taxa*, which have similar characteristics and have a common ancestry. The lowest taxonomic rank is the species. The next highest rank is the genus, followed in increasingly higher rank by family, order, class, phylum, and kingdom (fig. 35). No one knows how many species live on Earth today, or how many have existed throughout geologic time. We can, however, make some reasonable estimates. The range of living species is roughly 10 million to 100 million, though the number of modern plants, animals, and microbes actually documented is only about 1.4 million. Among this known total, 751,000 are insects and 248,000 are higher plants. That leaves only 416,200 for everything else—protozoa, algae, bacteria, viruses, and higher animals, including 4,000 mammal species.

It is easy to see, then, that extinction of a higher taxonomic category, such as a phylum, would limit the continuity of genetic lineages more severely than loss of a lower-level taxon, such as a species. Take, for example, our modern world. Tropical rain forests contain half or more of the world's plant and animal species. But each year, 27,000 of those rain-forest species are doomed to extinction, according to estimates by Edward O. Wilson, eminent evolutionary biologist and Curator in Entomology at Harvard's Museum of Comparative Zoology. Yet you and I don't notice affects from any of them. Let's suppose, however, that the entire phylum Craniata (fish, reptiles, amphibians, mammals, birds) were suddenly wiped out. It would be a catastrophic blow to the world's ecosystems, and would alter the evolution of higher organic life on Earth forever. (Actually, we wouldn't notice this either, since we would be among the missing.)

Most mass extinctions involve fewer than 20 percent of Earth's plant and animal species, but the five largest extinctions decimated as much as 50–95 percent of the species living at those times. The earliest of the "Big Five" took place 438 million years ago, and marks the boundary between the Ordovician and Silurian Periods (fig. 33). The second oldest major extinction was near the end of the Devonian Period, approximately 360 million years ago; the third, and most drastic, marks the end of the Permian Period, 245 million years ago. In the Mesozoic Era, one major extinction took place at the Triassic-Jurassic boundary (208 million years ago), and

Taxonomic Classification of Domestic Dogs

Taxonomic Level	Name	Examples
Kingdom	Animalia	All animals
Phylum	Craniata	Fish, reptiles, amphibians, mammals, birds
Subphylum	Vertebrata	Bony fish, reptiles, birds, amphibians, mammals
Class	Mammalia	Rabbits, deer, bats, mice, squirrels, apes, monkeys, cows, pigs, horses, cats, dogs, kangaroos, whales, elephants, humans
Order	Carnivora	Dogs, cats, bears, otters
Family	Canidae	Dogs, wolves
Genus	Canis	Dogs
Species	Canis familiaris	Domestic dogs

35. Scientists group organisms according to a formal classification that categorizes taxa according to their similarity of body features and ancestry. The classification starts with species at the lowest level and ends with kingdom, the highest level. Examples from the Kingdom Animalia are shown here. The other kingdoms are Bacteria, Protoctista, Fungi, and Plantae.

the most famous of all, marked by the demise of the dinosaurs, ended the Cretaceous Period, 65 million years ago.

The oldest and second-largest Big Five mass extinction took place at the end of the Ordovician Period. This event is noted for drastic reductions in the brachiopod and trilobite populations. Altogether, around 70 percent of all species disappeared. Late Ordovician environments were stressed by the buildup of continental ice sheets during the last few millions of years of the period. Sea level dropped as a consequence, and the vast, shallow, Ordovician seas dried up, devastating their biota. The ice sheets melted by the end of Ordovician time, however, allowing shallow marine waters once again to flood the continents. Life renewed itself in the early Silurian Period.

A few million years before the close of the Devonian, the second-oldest Big Five mass extinction further decimated the brachiopods (fig. 33). All the remaining trilobites disappeared, except for one small group. In addition, major groups of molluscs, corals, and fish were snuffed out for good. The total loss amounted to 35–40 percent of the biota. This extinction took place as the shifting North American and European continental plates joined near the equator. The joining closed off north-south circulation of warm tropical waters, contributing to a change in the global climate.

The Mother of All Mass Extinctions nearly eliminated all of Earth's inhabitants, other than microbes, at the end of the Permian Period (fig. 33). According to most estimates, 80–95 percent of all species disappeared. At higher taxonomic levels, 75 percent of all the reptile and amphibian families died out. By some estimates nearly 99 percent of all tetrapods were lost. In the sea, the great die-off was even worse; only about 4 percent of marine species survived the crash. Many marine families experienced 75–98 percent loss, and many others disappeared altogether.

In a geological instant, Earth's living communities were yanked from familiar evolutionary pathways they had followed for 300 million years, and diverted onto new, untraveled courses. To the good fortune of our mammalian lineage, however, one or two families of mammal-like reptiles survived the awful decimation.

The late Permian extinction took place at a time when all continental plates drifted together to form *Pangaea*, a single, gigantic supercontinent. Pangaea was nearly all dry land, with few, if any, shallow seas or large lakes. The continental interior was so far from any oceanic source of moisture, that it turned into a vast desert. Such a harsh environment allowed only a few hardy species to populate this vast central region. At the same time, glaciation at the poles withdrew water from the ocean and, thereby, lowered sea level. A low sea level, in turn, eliminated shallow-water habitats formerly available on continental shelves. The drained continental shelves exposed vast new tracts of land to weathering. Physical and chemical processes changed the composition of the surface rocks and produced soils. Some researchers have concluded that part of this weathering process generated unprecedented volumes of carbon dioxide from exposed coal beds. The CO_2 expansion might have reduced the amount of free oxygen in the atmosphere by as much as 50 percent. Such a severe loss of atmospheric oxygen would have reduced the ocean's share of oxygen as well, perhaps by as much as 80 percent. If true, this worldwide oxygen deficit would have slowly suffocated Earth's multitude of oxygen breathers.

Volcanic activity also accelerated dramatically during the late Permian. A massive outpouring of dark, fine-grained lavas flooded more than half a million square miles of Siberia. The lava flood continued for 600,000 years, until it formed a blanket 10,000 feet thick. Such unabated volcanism must have drastically altered global climate by spewing huge volumes of CO_2 and sulfur dioxide into the atmosphere.

Organisms that survived wholesale liquidation at the end of the Permian Period found it hard to recover their former places in the global community. The fact that life could recover from such a blow, expand again into all the vacant ecological niches, and produce the magnificent reptilian faunas of the Mesozoic Era, is one of the wonders of evolution's extraordinary weeding-out process. So far, natural selection has always managed to prepare at least one or two species for almost any imaginable environmental change. But the Permian crash also reminds us that our ancestral lifeline is scarily tenuous; the next episode of mass extinction might prove the exception to the rule.

Following the Permian devastation, early Triassic populations were weird. For roughly the first 5 million years of Triassic time, no large animals existed at all. Our hypothetical time-traveler would have been aghast at the sparsity of species inhabiting this strange land. Among the early Triassic tetrapods, 95 percent belonged to a single mammal-like reptile called *Lystrosaurus* (fig. 34). These three-foot-long, snub-nosed, pig-like plant eaters dominated terrestrial faunas on all continents, and apparently had no predators; they were virtually everywhere.

Among the other few successful early Triassic groups were some mammal-like reptiles. These ancient predecessors of mankind proliferated into a variety of plant-eating and meat-eating versions. Their evolutionary transition into mammals accelerated, so that by late Triassic time, they were difficult to differentiate from true mammals. Other surviving groups were just getting into high gear at the end of Triassic time, 40 million years after the Permian crash. At that point, as luck would have it, all were fatally hammered by the fourth Big Five extinction event (fig. 33).

The terminal Triassic extinction may actually have been two closely spaced events. The earlier event, about 20 million years before the end of the Triassic, wiped out mainly land faunas, including many primitive reptile lineages. But it opened up new niches to the dinosaurs, which seized their opportunity and left all challengers in the dust for the next 150 million years. The second extinction terminated the Triassic Period and ravaged mainly marine animals. Reef communities, fish, sea urchins, scallops, and others suffered major declines.

The end-of-Triassic extinctions took place as Pangaea began to break up again into the continental plates we recognize today. Also, the Atlantic Ocean began to form. It was initially a long narrow seaway, which brought much-needed moisture into the heart of Pangaea's vast central desert.

The prolific reptile faunas, especially the dinosaurs, which characterized the rest of the Mesozoic Era (Jurassic and Cretaceous Periods) are well known to many readers. They have been globally publicized in books, videos, movies, and museums, and by all other mass communication media, ever since the legendary *Iguanodon* was publicly displayed in London's Crystal Palace back in 1851. In their late Mesozoic heyday, reptiles literally took over the world. From sky to continent to ocean depth, from pole to pole, and from gigantic, plodding waders to tiny, upright racers, reptiles did it all. But even these impressive, exceedingly successful creatures could not escape the next flick of evolution's fickle finger—the Cretaceous-Tertiary extinction, or K-T event (fig. 33). (It may seem odd that this is not called the C-T event, but geologists use "C" as shorthand for Carboniferous. The "K" comes from *Kreide*, the German word for chalk, which is an abundant rock type in many Cretaceous deposits.)

The K-T event is the most publicized and thoroughly studied mass extinction event we know about. Even so, interpretations of its cause and effects are varied and controversial. Reputations have been made and ruined, and friendships have been formed and lost, in public disputes over the K-T event. The principal debates concern: (1) the time span of the event (Was it virtually instantaneous, or did it proceed stepwise over several millions of years?); and (2) the principal cause of the event (Was it a slow climatic deterioration, a sea-level change, a massive volcanic outpouring, or a giant meteorite impact?).

As more and more data have been acquired, it seems clear to me that this extinction was a highly complex process, and that all of the above are more or less true. The late Cretaceous environments began deteriorating over a span of several million years, and several lesser stepwise extinctions took place. But there is much compelling evidence that these slower biotic changes were topped off by instantaneous changes—sea level fell, and the circulation patterns, chemistry, and temperature of the ocean changed significantly. In fact, many have argued that the impact of a six-mile-wide meteorite could have produced all these rapid changes and more. The impact itself would have lasted only a few minutes, but its short-term aftereffects, such as giant tsunamis, may have lasted for weeks. The impact also would have set off a succession of long-term atmospheric and oceanic disturbances, which could last for a few tens of thousands of years. I will discuss these more fully in chapter 10.

After extensive studies of the K-T boundary extinctions, it is not surprising that meteorite evidence has turned up at many other (but not all) times of mass extinction, including the remaining four of the Big Five. An iridium spike has been found, for example, at or near all of the Big Five extinctions. Microtektites and/or shocked minerals have been reported from four of the Big Five (not yet from the Ordovician-Silurian boundary). So far, however, only the K-T event, among the Big Five, has a proven source crater. Possible candidates for some of the others exist (end of Permian and end of Triassic, for example), but need further study to pinpoint their ages.

Meanwhile, continuing research is turning up more and more abrupt changes in the geochemistry of rocks at many extinction boundaries. All the Big Five extinction events are marked by sudden, short-lived shifts in the ratios between different isotopes of oxygen, carbon, sulfur, and strontium. These shifts are additional evidence for the kind of instantaneous global disruptions produced by meteorite impacts. In fact, meteorite impacts can account for all the biotic, geochemical, climatic, and sea-level changes blamed for the Big Five extinction events. The way things are going, we shouldn't be surprised if twenty years from now, all the arguments to the contrary will seem naive. But is the converse true? Do all large meteorite impacts cause mass extinctions? We will explore that question further in chapter 10.

Despite all the additional research, the most thoroughly understood of all mass extinctions is still the K-T event. Scientists have confirmed abrupt worldwide extinctions among a host of marine and terrestrial species, genera, and families, ranging from forams to dinosaurs. All took place at, or very close to, the K-T boundary, and only about 35 percent of the Cretaceous organisms crossed the fossil-free, iridium-rich boundary clay. The earliest Cenozoic biota were impoverished; the number of species was greatly reduced, and the reptiles ruled no more. It was time for the mammals to shine, and shine they did. By Eocene time, they had taken command of most of the "professions" practiced by their former reptile competitors, and they flourished from coastal plain to mountainside. The oceans, too, claimed a significant community of Eocene mammals, in the form of primitive whales.

CHAPTER TEN

Eocene Environments

THE Eocene Epoch dawned approximately 55 million years ago, ten million years after the dinosaur mass extinction. Most geologists subdivide the epoch into three parts—early, middle, and late. The part of most interest to readers of this book is the late Eocene, which spanned the interval 37 million to 33.7 million years ago. The Eocene was preceded by the warmer Paleocene Epoch and followed by the colder Oligocene (fig. 36).

If you had been there in the early Eocene, you would not have recognized North America. It was a hot, humid continent, covered by tropical rain forests, like those of modern Central America. Vast fetid swamps collected immense piles of rotting vegetation. We see them still preserved today as seams of coal, some as thick as three hundred feet. Even the Eocene polar regions supported lush plant life and animals whose modern relatives inhabit warmer latitudes.

This tropical climate wasn't caused by the North American tectonic plate being farther south than it is today. By Eocene time, the shape and distribution of continents was much like the modern configuration. The entire Eocene Earth was a much warmer greenhouse world, populated by warmth-loving plants and animals.

Any nature lover easily would have recognized a few of the early Eocene animals. No large dinosaurian reptiles remained, but modern-looking crocodiles, turtles, frogs, and salamanders thrived in the steaming tropical swamps and bogs. Most of the early Eocene inhabitants, however, were strange to behold (fig. 37). Eight-foot-tall, flesh-eating flightless birds, called *Diatryma*, stalked the forest underbrush as the dominant predators. Though carnivorous mammals were common, the largest, called *creodonts*, were no bigger than wolves. Most were diminutive, the size of domestic cats.

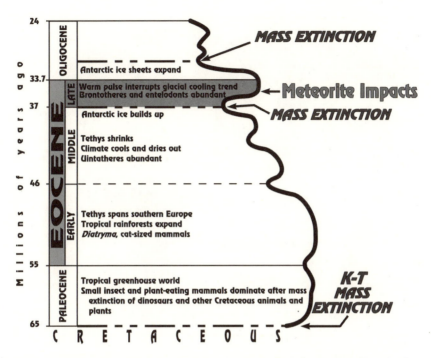

24

OLIGOCENE

33.7

Antarctic ice sheets expand

MASS EXTINCTION

Warm pulse interrupts glacial cooling trend
Brontotheres and entelodonts abundant

Meteorite Impacts

37

LATE

Antarctic ice builds up

MASS EXTINCTION

MIDDLE

Tethys shrinks
Climate cools and dries out
Uintatheres abundant

EOCENE

46

EARLY

Tethys spans southern Europe
Tropical rainforests expand
Diatryma, cat-sized mammals

Millions of years ago

55

PALEOCENE

Tropical greenhouse world
Small insect and plant-eating mammals dominate after mass
  extinction of dinosaurs and other Cretaceous animals and
  plants

**K-T
MASS
EXTINCTION**

65

C R E T A C E O U S

36. A geologic time chart showing the three parts of the Eocene Epoch, the adjacent epochs, and some of the main biotic, climatic, and smaller mass extinction events that succeeded the final Big Five mass extinction at the Cretaceous-Tertiary (K-T) boundary. The paleotemperature curve *(heavy lobate line)* shows a step-like cooling trend (cooler to the left) starting in the early Eocene and ending in the Oligocene. A distinct warm pulse interrupted the cooling trend at the same time the three late Eocene meteorites struck.

The prey of these meat-eaters were mainly primitive hoofed mammals that scurried about on the floor of the rain forests. Sausage-bodied *hypsodonts* and sheep-sized *phenacodonts* were most abundant. Herds of *Hyracotherium*, the earliest horses, also populated the rain-forest floor, but you could not have ridden one. They were no larger than a dachshund. You would have been startled to see their feet; each front foot was supported by four dainty, hoofed toes, and each rear foot had three. These "dawn horses" lived side by side with small tapir-like and rhino-like hoofed mammals. One of the oddest was the cow-sized *Coryphodon*. Its patchwork body had a huge head, a pig-like snout, sharp tiger-like ca-

37. Four representative mammals and a giant predatory bird, which populated the formally recognized subdivisions of the Eocene Epoch.

nine teeth, and broad hooves, comparable to those of a cow. The *tillodonts* and *taeniodonts* were even more bizarre. They resembled bear-sized beavers, with huge, chisel-like incisors.

Up in the trees thrived imposing populations of lemur-like primates, along with *multituberculates*, which resembled squirrels. Some of the primates had keen eyes, opposable thumbs, and long tails, but their brains were typically smaller than modern counterparts. The multituberculates, named for their knobby molar teeth, had chisel-edged, squirrel-like incisors, but also grew prehensile tails, like some modern monkeys. Because of their incisors, multituberculates are considered to have been odd Eocene equivalents of rodents. Most of these tree-dwelling animals ate leaves and fruit, which were abundant in the lush tropical forests.

Though the configuration of Eocene and modern continents was similar, the early Eocene oceans were strikingly different from today's. The greatest difference was the presence of a broad seaway, named *Tethys*, which connected the Atlantic to the Pacific right across southern Europe. This broad belt of tropical water girdled the equator from the Strait of Gibraltar to Indonesia. Warm Tethys waters strongly influenced the distribution of atmospheric heat and the circulation of Eocene ocean currents.

Also, the early Eocene sea level was much higher than today. Vast areas of the continental margins were inundated by marine waters. The organisms inhabiting early Eocene oceans would have been more familiar to us than most of their terrestrial cousins. A wide variety of modern-looking clams and snails dominated seafloor habitats. There were also hoards of sea urchins, starfish, and sand dollars. Coral reefs were widespread in the shallow warm seas, especially along the margins of Tethys. Their fossil remains can be observed today in outcrops in Italy, Spain, and other Mediterranean localities.

Microscopic plants and animals flourished in nutrient-rich water masses near the ocean surface, and also in diverse environments on the ocean floor. Some unusually large bottom-dwelling foraminifera, called *Nummulites,* were so abundant that their remains formed massive limestone deposits. The ancient Egyptians quarried the nummulitic limestone to construct their famous pyramids.

Marine vertebrates also would have been easily identified. Diverse schools of bony fish resembled modern species, and were prey to familiar-looking sharks. Even primitive whales roamed the early Eocene oceans.

The early Eocene Earth, then, embraced a lush tropical Eden whose warmth reached even to the poles. The excess Eocene heat is thought to have been the result of elevated levels of so-called *greenhouse gases*, particularly carbon dioxide. In Earth's atmosphere, the layer of carbon dioxide acted like a two-way mirror for heat. It transmitted the heat of sunlight *inward* to the Earth's surface, but it was a barrier to heat radiated *outward* from the Earth's surface. It trapped heat like the glass panes of a nursery greenhouse.

This unusually warm, moist climate lasted for 15 million years of early and middle Eocene time. During the middle Eocene, however, global temperatures began to drop. Rainfall diminished, and ice sheets began a slow buildup in Antarctica. By the end of the middle Eocene, the Tethyan seaway was only a narrow remnant of its former expanse. It no longer effectively distributed tropical heat along the equator. It didn't take long for higher latitudes to cool down and become temperate climate zones, more like today.

Middle Eocene terrestrial plants and animals, likewise, began to change. On the North American continent, tropical rain forests were replaced by open woodlands of deciduous trees, which were adapted to the cooler, drier climate. The climate shift, in fact, reduced the area of tropical rain forests over the entire globe. Terrestrial animal populations also responded quickly to the cooler, drier climate. The total number of species (diversity) dropped, but many new species arose in major spurts of evolutionary change.

Perhaps the most spectacular middle Eocene mammals were the elephant-sized *uintatheres*, which bore three pairs of knobby horns and huge canine tusks (fig. 37). By the end of the middle Eocene, terrestrial animals in North America experienced a dramatic reduction in the number of species, including the uintatheres. In some areas, more than 80 percent of middle Eocene land animals failed to survive into the late Eocene. The extinctions did not extend worldwide, however. European vertebrate species underwent no unusual reductions at the end of the middle Eocene.

Marine communities also diminished abruptly at the close of the middle Eocene. The variety of different species was much reduced, and many species went extinct. In composite, though, the marine changes generally were not as drastic as those taking place on the continents.

Thus, by the time the late Eocene began, 37 million years ago, Earth had developed a cooler, drier climate, and cooler oceans than the preceding two-thirds of the epoch. Still, it was a much warmer and wetter world than today. Sea level remained substantially higher, as well. On a geological timescale, however, global climate can shift quickly and often, as we saw in chapters 8 and 9; late Eocene climate remained relatively cool and dry for only 2 million years.

At about 35 million years ago, Earth's climate switched once more to a warmer mode. The oceans warmed up by about 2°F, as indicated by the ratio of heavy to light isotopes of oxygen present

in the shells of late Eocene microfossils. Geochemists count isotopes with a specialized instrument known as a mass spectrometer.
Inferring ancient ocean temperatures from isotopic ratios is akin
to estimating seasonal temperatures from the ratio between fat
cows (= heavy isotopes) and thin cows (= light isotopes) in a large
herd. Assume that more fat cows are present in winter than in
summer. With only a photograph of the herd taken thirty years
ago you could determine whether the season had been warm or
cold.

Even though the late Eocene climate warmed, marine organisms
did not return to their middle Eocene abundances. Their principal
response to the warming was a marked migration of warm-water
species into higher latitudes. On land, on the other hand, both
plants and animals responded to this warm shift more dramatically.
Documented changes in terrestrial plants suggest that temperatures over North America increased by 11–14°F in a few million
years. Rain forests reclaimed many of their abandoned higher-lati-
tude tracts. Through those verdant forests roamed massive *brontotheres*, even larger than the uintatheres, each with a pair of huge blunt
nose-horns (fig. 37). Along with the brontotheres were species left
over from the middle Eocene, including the primitive primates,
multituberculates, and primitive tapirs. Crocodiles and turtles also
still abounded. On the other hand, many new groups of mammals
evolved in North America, and others migrated from Europe, including *Archaeotherium*, a giant, pig-like entelodont.

The general aspect of late Eocene mammal populations was de-
cidedly modern. The first true squirrels, beavers, pocket gophers,
pocket mice, peccaries, and pangolins lived among herds of
horses, camels, and rhinos. Rabbits were particularly abundant.
Most of these semi-modern mammals were relatively small—rarely
larger than a German Shepherd dog.

We might say then, that 35 million years ago the world was in
transition. The biosphere had experienced a major extinction
event two million years earlier at the end of the middle Eocene,
which coincided with a distinct shift to cooler, drier continental
climates, and cooler oceans. But the fossil record of continental
and marine organisms indicates a subsequent return to warmer,

more moist paleoenvironments at 35 million years ago. That date of 35 million years has a familiar ring, doesn't it? It's that magic number, the age of the Chesapeake Bay meteorite impact. Could the impact have been responsible (or partly responsible) for the pulse of climatic warming? If not, what can we say about the impact's regional and global effects?

A key ingredient controlling the environmental damage from a meteorite impact is the amount of very fine dust particles blasted out into the Earth's atmosphere. Theoretical calculations derived from nuclear weapons tests indicate that a meteorite 2–3 miles in diameter would have been required to produce a crater as large as that of Chesapeake Bay. Such a large impactor would yield a burst of kinetic energy equivalent to approximately 10 trillion tons (10 million megatons) of TNT. This violent release of energy produces mind-boggling environmental destruction. For comparison, the potential energy yield from the world's entire nuclear arsenal is equivalent to around 100 billion tons (100 thousand megatons) of TNT.

As I mentioned earlier, the initial product of the meteorite blast would have been a planet-rocking shock wave that propagated downward and outward from ground zero at supersonic speed. In contrast to this megashock, all historical earthquakes have been mere hiccups. The ground shock would have been followed in a matter of seconds by a scorching atmospheric blast wave, hot enough to incinerate every living thing it touched. Then an enormous vapor plume would rise high into the atmosphere. The vapor plume included the vaporized meteorite itself and a vast volume of vaporized target rocks and ocean water. The plume would have expanded so rapidly that part of its energy was converted into a sizzling-hot super-hurricane or *hypercane,* raging at more than 500 miles per hour through the lower atmosphere. Into this maelstrom was ejected millions of tons of white-hot particles of pulverized rock and dust. Most of the larger particles would cool and fall back to the surface within a day or two. As they hurtled back to Earth, atmospheric friction would have reheated them into glowing embers, which ignited forest fires all over the globe. Additional wildfires, hundreds of miles across, would be ignited by thermal radiation emitted by the shock-roasted atmosphere.

The enormous volume of fine dust thrown into the atmosphere would blanket the entire planet within a few weeks, and remain suspended there for several months. To this murky blanket was added smoke and ash given off by the roaring wildfires. All this airborne dust retarded the passage of sunlight to the Earth's surface. As a result, plants were unable to maintain photosynthesis, and rapidly died off. Global darkness also would have limited the vision of most higher animals and hindered their foraging for food. In addition, the loss of solar radiation would cool down the surface of the continents enough to create a deep, prolonged winter, which would destroy temperature-sensitive plants and animal life on a horrific scale.

In the case of an oceanic target, as at Chesapeake Bay, the volume of impact-generated water vapor injected into the atmosphere would be especially immense. This excess vapor alone would drastically change atmospheric chemistry and the distribution of global heat. The profusion of hydrogen and chlorine molecules (from vaporized sea water) injected into the atmosphere, for example, would severely diminish the protective ozone layer and intensify carcinogenic ultraviolet radiation to the surface.

Atmospheric shock waves produced by the meteorite's passage would cause nitrogen and oxygen molecules to combine, thereby generating prodigious volumes of nitric oxide. The nitric oxide then would combine with water vapor to form nitric acid, which, in turn, would lead to corrosive acid rainfall on a massive scale. Meanwhile, the East Coast of North America was being pounded by the largest tsunami waves produced in the last 35 million years. Life on Earth would have been shocked, vaporized, pulverized, barbequed, blinded, irradiated, acidified, drowned, starved, and frozen—Eocene Eden literally "went to hell."

You will not be surprised, then, that some investigators predict that a meteorite impact large enough to produce the Chesapeake Bay crater should have extinguished 45 percent of Earth's marine species. On the other hand, you might be astonished to learn that, so far, geologists and paleontologists have failed to find clear evidence of such devastation. It is not for lack of searching, either.

Scientists have collaborated in a concerted effort during the last ten years to analyze beds of late Eocene age all over the globe. I have incorporated their results into preceding paragraphs of this book. To date, however, they have discovered no beds containing masses of instantly barbequed late Eocene organisms, or other signs of mass extinction events in 35-million-year-old sediments.

The apparent lack of late Eocene extinctions is even more remarkable when we consider that at least two additional meteorites struck Earth 35 million years ago. One excavated the small Toms Canyon crater described earlier in chapter 4. The other created Popigai crater, located on the northeastern slope of the Anabar shield in Northern Siberia, near the shore of the Laptev Sea (fig. 38). Popigai is identical in size to the Chesapeake Bay crater, about fifty miles in diameter. Like Chesapeake Bay, Popigai is a complex peak-ring crater. In contrast, however, Popigai is exposed at the surface. Geologists have studied it extensively from outcrops and cores. Also in contrast to the Chesapeake Bay impact, the Popigai meteorite struck a high-latitude continental site. It was not covered by ocean waters. The thickness of the pre-impact sedimentary column (about 0.6 mile) is similar to that of Chesapeake Bay, but the Siberian rocks are much older—mainly Proterozoic to Triassic, 700–250 million years old. These old rocks are highly consolidated quartzites, *dolomites*, limestones, sandstones, and *argillites*. The thick limestones and dolomites at Popigai, all made mainly of calcium, carbon, and oxygen, would have been excellent sources for additional carbon dioxide, with which the impact could saturate the atmosphere. Beneath the sedimentary beds at Popigai, as at Chesapeake Bay, is a crystalline basement composed mainly of ancient metamorphic rocks (gneisses, *schists*) and granites, more than 600 million years old.

It is quite plausible to presume that the Popigai, Chesapeake Bay, and Toms Canyon meteorites struck nearly simultaneously, or in rapid succession. At present, their ages are not clearly distinguishable from each other on the basis of any of the best dating techniques (fossils, magnetic reversals, isotope ratios, and nuclear fission-track analysis). Furthermore, the three respective meteorites appear to have struck at low angles to the horizon, all traveling

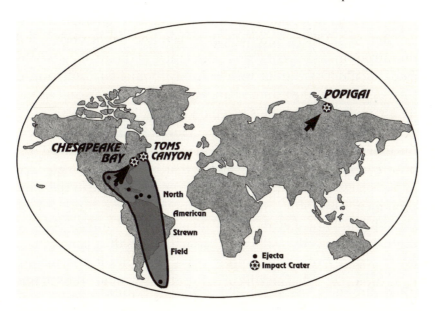

38. Modern configuration of continents shows the location of three known late Eocene impact craters, each of which is 35 million years old. Arrows depict the inferred trajectories of the three meteorites that excavated these craters. Tektites and shocked minerals of the North American tektite strewn field also are 35 million years old, and appear to have been derived from the Chesapeake Bay crater.

from southwest to northeast. This suggests an analogy to the consecutive impacts of comet Shoemaker-Levy 9 into Jupiter, which awed TV audiences during the summer of 1994.

Comet showers are thought to occur so commonly (on a geologic scale) in our solar system that some researchers believe *most* meteorite craters were excavated by paired or multiple impactors. Several examples of twin craters have been documented on the Earth, and linear chains of craters have been reported on other planets and moons. The most recent evidence for a late Eocene comet shower comes from isotope studies. Italian outcrops containing late Eocene shocked quartz and an iridium spike also are unusually enriched in the light isotope of helium (^{3}He). Such concentrations of helium-3 are thought to be derived from interplanetary dust particles incorporated in comets.

So, perhaps other late Eocene impact craters still await discovery. But even if there were only three late Eocene meteorites and they struck thousands of years apart, either of the two larger ones alone should have caused worldwide environmental disturbances. In combination, they would have equaled some estimates of the devastation caused by the Chicxulub impact.

We cannot avoid the fact that the known Eocene meteorite extinction record does not correspond with the expected model. We must consider, then, the possibility that not all giant meteorite impacts have caused massive species extinctions. One of the most prominent meteorite extinction hypotheses was proposed by David Raup, a noted evolutionary theorist at the University of Chicago. Raup created a "kill curve," which relates the diameter of a given impact crater to the percentage of species it blasted into extinction (fig. 39). The kill-curve hypothesis assumes that all meteorite impacts have caused mass extinctions. Other authors have suggested that the reciprocal also is true; that all mass extinctions have been caused by meteorite impacts. The lack of extinctions associated with the late Eocene impacts, however, nullifies both assumptions. Instead, current data suggest that if meteorites do trigger mass extinctions, it is only those that produce craters larger than fifty miles in diameter. Or perhaps other special geological or paleoenvironmental conditions have to be combined with the impact.

The lack of verified extinction events 35 million years ago does not rule out the possibility of other types of impact effects on the biosphere, however. There are tantalizing clues of distinct late Eocene paleoenvironmental alterations in some localities. For example, an outcrop of upper Eocene beds on the Caribbean island of Barbados, exposes a tektite-bearing layer the same age as the Chesapeake Bay impact. This layer is part of the North American tektite strewn field. At this layer, five species of silica-shelled microfossils known as *radiolarians*, disappeared, never to return to the Barbados locality. The fossil remains of those five radiolarians are abundant below the tektite bed, but none are present above it. In contrast, fossils of many other radiolarian species at the Barbados locality occur above and below the tektite layer without interrup-

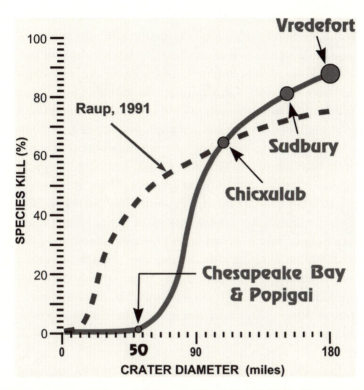

39. David Raup constructed a "kill curve" to relate the percentage of species extinction to the diameter of impact craters. Because no mass extinctions are associated with the 50-mile-wide Popigai and Chesapeake Bay craters, however, Raup's curve *(dashed line)* must be revised *(solid line)*. The Sudbury and Vredefort craters (see chapter 15) are the two largest documented terrestrial craters, and as such, define the high end of the kill curve.

tion of their lineages. At DSDP Site 612, off New Jersey, however, fossils of these same five radiolarian species are present well above the tektite layer. In other words, though five species disappeared at Barbados coincident with the meteorite impact, all five continued to occupy the 612 locality many thousands of years after the impact. The fossil record is full of similar local disappearances, most of which were caused by something other than a meteorite impact.

So far, there is no evidence that any other group of marine or terrestrial organisms underwent even local disappearances in outcrops or cores that contain meteorite deposits 35 million years old. What we detect, instead, are changes in the abundance of some species and alterations in their global distribution patterns. Even more important are changes in the skeletal chemistry of certain late Eocene fossils. For example, the ratios of two oxygen isotopes in the tests of foraminifera change markedly near the tektite layers. These types of changes are clues to ancient environmental shifts. In this case, the isotopic change indicates that a pulse of warm climate coincided with the Chesapeake Bay, Popigai, and Toms Canyon meteorite impacts (fig. 40). The warm pulse may indicate that the global climate responded to the triple meteorite strike, even though no coincident mass extinctions took place. The late Eocene fossil record seems to suggest that instead of an immediate mass extinction following the impacts, several small, successive extinction events took place, unrelated to the impacts. The smaller kills eventually were followed by a dramatic mass extinction, but that didn't take place until the early Oligocene, approximately one million years after the late Eocene warm spell ended.

In other words, the late Eocene appears to have undergone a delayed environmental and biotic response to a complex combination of triggering mechanisms. First a notable buildup of ice sheets in Antarctica coincided with climatic cooling in the middle Eocene. Then the triple meteorite strike 35 million years ago created a pulse of greenhouse warming in the midst of the long-term global cooling trend. The meteorite-generated warmth reversed the global cooling trend for a million years before the heat began to dissipate. When the warmth finally abated, Antarctic ice growth accelerated, and a particularly sharp global temperature drop followed. That is when the early Oligocene mass extinction took place. We can surmise that without the warm pulse, the falling global temperatures would have produced an extinction event a few million years earlier, in the late Eocene. Thus, atmospheric warming from the three meteorite strikes may actually have delayed, rather than caused, a mass extinction.

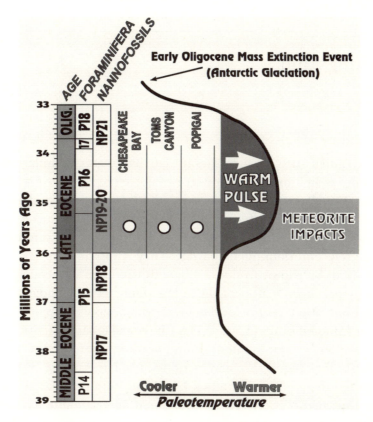

40. A pulse of warm climate coincided with the late Eocene meteorite impacts. The heavy curved line represents paleotemperature changes through time. In this figure, I have assumed simultaneous impacts, though they may have been separated by several tens of thousands of years. An early Oligocene mass extinction followed the termination of the warm pulse.

CHAPTER ELEVEN

Sinking Sand

DISCOVERIES such as the Chesapeake Bay crater stimulate the human fascination with extraterrestrial events. But you may ask, "Why should the federal government spend hundreds of thousands of dollars researching a buried impact crater?" The U.S. Congress, which funds federal research efforts, is not interested in merely entertaining taxpayers. They want practical results, which can be shown to directly benefit the nation's needs. This view is particularly noticeable today, when federal budgets are under increased scrutiny, and many departments and agencies are being trimmed or eliminated. The Chesapeake Bay meteorite impact and its resultant crater are not, however, merely exotic phenomena that pique the imagination. That late Eocene blast rearranged the geological framework of the Virginia Coastal Plain and fundamentally altered its subsequent development. Residual effects of the impact still influence the lives of citizens living around Chesapeake Bay, though they are separated from the crater by 35 million years of geological time and thousands of feet of sedimentary rocks. We will now examine both the ancient and modern consequences of that cosmic collision.

To begin with, the presence of the Chesapeake Bay crater explains several unusual aspects of coastal-plain geology that have puzzled local experts for decades. I have already explained that Cede Cederstrom's Eocene basin is actually part of the crater, the Hampton Roads fault is a section of the crater rim, and the Mattaponi Formation is mostly equivalent to the Exmore breccia, which fills and surrounds the crater. Cederstrom and later investigators also determined that the Chickahominy Formation, a 300-ft-thick clay bed that caps the Exmore breccia, is confined to the subsurface. Nowhere can the Chickahominy be found in ground-level

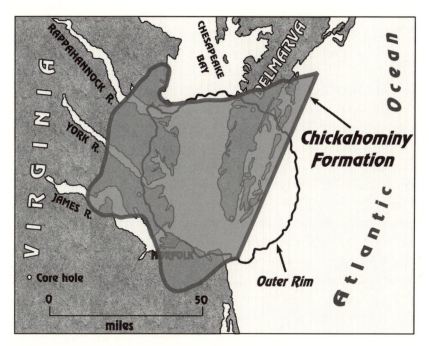

41. The distribution of the late Eocene Chickahominy Formation, which rests on top of the Exmore breccia, nearly matches the outline of the Chesapeake Bay crater.

outcrops. Moreover, when USGS geohydrologist Phil Brown and his colleagues completed comprehensive ground-water studies in the early 1970s, they found the Chickahominy Formation to be distributed in a crescent-shaped band arcing across the western margin of the bay. When I updated Brown's map, using data from the seismic profiles, the Chickahominy plotted almost entirely inside the crater's outer rim (fig. 41). The formation is also significantly thicker over the crater than outside it. Clearly, both its distribution and thickness are somehow related to the presence of the crater—but how?

Despite its restricted areal distribution, the Chickahominy originally must have been spread over a much larger area. I infer this from the relatively deep water in which it accumulated. By analogy with modern foraminiferal assemblages, the Chickahominy foraminifera would have occupied water depths of six hundred feet or more. That means that the Eocene Atlantic shoreline was much farther west than it is today; it pressed against the eastern flanks of the Appalachian Mountains. The Chickahominy Formation undoubtedly also extended much farther west, and blanketed the seafloor over much of what is now southeastern Virginia. A few million years after the Chickahominy accumulated, however, in the early part of the Oligocene Epoch, the sea receded to an exceptionally low level as vast ice sheets rapidly expanded over Antarctica. The retreating shoreline exposed much of the shallow Atlantic seafloor, which then became part of the Virginia Coastal Plain. Upon exposure, the Chickahominy Formation eroded extensively. On higher elevations, it was almost completely removed or significantly thinned. Only in the broad depression over the crater was the Chickahominy preserved in its original thickness. The modern distribution pattern of the Chickahominy, therefore, was determined mainly by the crater's ability to preserve it. Without the crater, there would be no more Chickahominy Formation, and the late Eocene chapter of Virginia's geological record would contain only blank pages.

Geologists studying the Virginia Coastal Plain have long held that the Earth's crust south of Norfolk has been uplifted relative to surrounding areas. Sedimentary beds younger than Eocene in age are unexpectedly thin and are at higher elevations south of Norfolk than at localities to the north. This alleged uplift has been described in geological literature as the *Norfolk uplift* or *Norfolk arch* (fig. 42). But now we can propose a different cause for the higher elevations and thinner beds. Both are due to the crater's presence. Instead of being an isolated uplift, the "Norfolk arch" is merely part of the rim of the crater. The beds are notably thicker and at lower elevations to the north, because they are inside the crater where the underlying Exmore breccia has been compacting and subsiding. These two processes provided a deeper post-Eocene basin north of Norfolk, which could hold a thicker pile of Oligocene to Quaternary sediments. In fact, long-term subsidence of the Exmore breccia is the mechanism that has transmitted geologic, geographic, and environmental effects of the 35-mil-

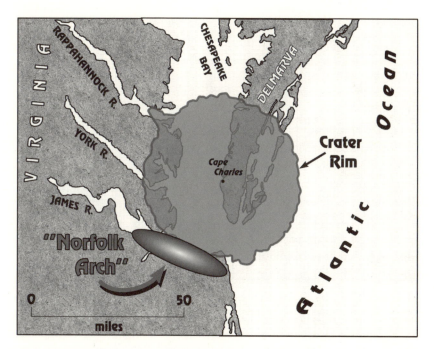

42. The "Norfolk arch" was originally envisioned as an upwarp on the surface of the crystalline basement. We now know that the so-called arch is part of the outer rim of the Chesapeake Bay crater.

lion-year-old meteorite impact to the modern Chesapeake Bay region.

The most populous urban corridor in Virginia encompasses more than 1.5 million people in the cities of Virginia Beach, Norfolk, Portsmouth, Chesapeake, Newport News, and Hampton. These population centers are particularly susceptible to subsidence effects, because they overlie the southwest rim of the crater. Even more strongly affected is the Delmarva Peninsula, because it extends right out over the middle of the crater, the region of maximum subsidence. How did this subsidence mechanism maintain itself through 35 million years?

Following the impact, successive deposition of the Chickahominy Formation and younger sedimentary beds began to compact the unconsolidated, water-saturated Exmore breccia, like stacking bricks on a wet sponge. The weight of overlying sediments in-

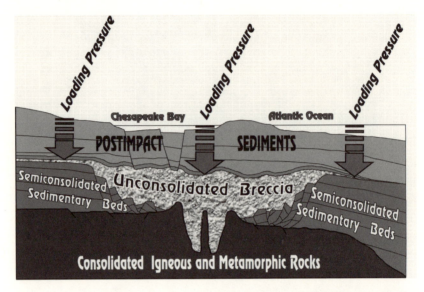

43. Differential subsidence of the land surface over the Chesapeake Bay impact crater is caused by rapid compaction of the unconsolidated Exmore breccia relative to slower compaction of semiconsolidated sedimentary beds surrounding the crater. The load of postimpact sediments provides the compaction force.

creased steadily through time as new layers were added. This loading pressure caused the surface of the breccia to subside more rapidly than the semiconsolidated sediments outside the crater; this is called *differential subsidence* (fig. 43). As differential subsidence progressed, the seafloor sagged, and a bowl-shaped depression formed over the crater. This, in turn, caused postimpact sedimentary beds to thicken over the crater, as they tried to fill in the depression. This depression and its thickened formations are among the most obvious features displayed on the seismic profiles.

It is not surprising that the breccia compacted and subsided. These processes are normal steps in the deposition and preservation of sedimentary formations. What *is* surprising is that the subsidence depression can be detected at ground level today. We see evidence of it in the geological and topographical features of the modern land surface around Chesapeake Bay. For example, sedimentary beds that form the existing ground surface inside the crater rim are geologically younger than surface beds outside the

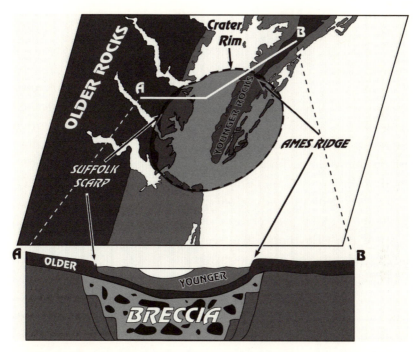

44. The distribution of geologic formations and topographic features of the present land surface reflect the presence of the underlying Chesapeake Bay impact crater. Greater differential subsidence of the land surface and bay floor is driven by compaction and subsidence of the buried Exmore breccia.

crater. We can follow the contact between the older and younger surface beds as an arcuate boundary, which curves around the western margin of the bay. The boundary traces the western arc of the crater rim (fig. 44).

The elevation of the ground surface also changes sharply along this arcuate geological boundary. Outside the crater, the ground elevation averages 30–45 feet higher than inside the crater rim. The elevation changes abruptly near the rim and creates a curved steep slope known as the *Suffolk Scarp*. If you drive through the tidewater region of Virginia along Route 17 from Newport News to Gloucester, you travel right along a flat surface called the Newport News terrace. The Suffolk Scarp forms the eastern boundary of this terrace, about a mile east of the highway. You are right on top of the crater rim. One of the best places to examine the scarp itself is along Virginia State Highway 14, between Gloucester and the village of James Store.

A similar geologic and topographic expression of the crater rim can be observed on the Delmarva Peninsula. Take U.S. Highway 13 north from Kiptopeke to Painter. A sharp, 15-ft rise in topographic elevation to the west of the road near Painter is called Ames Ridge. The ridge approximates the position of the buried crater rim in this area. Ames Ridge, like the Suffolk Scarp, also marks the geological boundary between younger sedimentary formations inside the crater and older sediments outside the crater.

These geological and topographical features are not the only evidence of the buried crater's residual effects. If you examine a map of the Chesapeake Bay region, you will notice something strange about the courses of two major rivers that enter the bay. The Rappahannock and other rivers north of the crater take a nearly direct southeasterly route to the bay. But the James River takes a quite different route. As it approaches the bay, the James makes a sharp right-angle bend to the northeast just as it crosses the buried crater rim, and flows directly toward the crater center (fig. 45). The York River does the same thing. These two rivers are the only modern ones that cross the crater rim. The lower course of each appears to be controlled by the crater's location, though its rim is buried more than one thousand feet below. The river diversions are additional results of the long-term differential compaction and subsidence of the Exmore breccia.

Similar control of older, ice-age river courses can be inferred from a different series of seismic reflection profiles, which image only the upper three hundred feet or so of the bay floor. Steve Colman, another of my Woods Hole colleagues, is an expert on Quaternary geology and modern landforms. Steve has mapped the youngest beds in the Chesapeake Bay in great detail, and has collaborated with Bob Mixon, who studied the same young beds on the Delmarva Peninsula. Mixon sampled the Delmarva formations by drilling short cores at many different sites. Together, these researchers mapped the courses of three ice-age river channels buried beneath the bay and Delmarva Peninsula. Like the modern

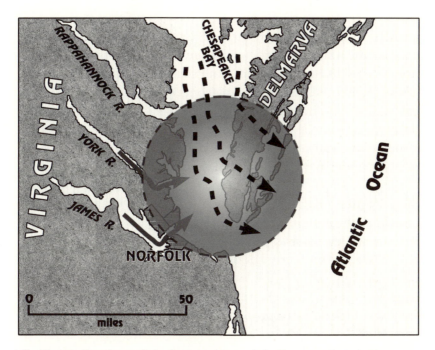

45. Abrupt diversions of the lower courses of the James and York Rivers are caused by differential subsidence of the Exmore breccia within the Chesapeake Bay impact crater. The heavy dashed lines represent courses of three buried ice-age channels (decreasing in age from north to south) eroded by the ancient Susquehanna River. The courses of these ice-age channels also were diverted sharply by the structural sag over the subsiding breccia.

James and York, these ancient rivers also abruptly changed course upon crossing the buried crater rim.

Such abrupt changes in topographic elevations and river courses seldom take place without a good geological reason. They are additional evidence that the 35-million-year-old Chesapeake Bay crater still influences the lower region of Chesapeake Bay today. They indicate that the land surfaces, bay bottom, and seafloor inside the crater rim have continued to subside faster than areas outside the rim for nearly 35 million years, at least up to a few thousand years ago, when the rivers cut their modern channels.

There is one more interesting example of how differential subsidence over the crater has continued right up to the present. In the last few years, much attention has been focused on major changes in our global environment. Such geological processes as greenhouse warming, sea-level rise, and glacial melting are weekly topics of discussion in scientific journals and popular news media. These phenomena are of growing concern to economic and political leaders, who have convened two international conferences to debate ways to cope with their effects. Potential threats to human populations from these global changes have spurred a renewed effort to understand, predict, and modify or control them. As part of the research effort, scientists and engineers are monitoring and modeling the relative rate at which sea level is rising along many coastlines throughout the world. The term "relative" here means that the sea level is rising relative to the land surface. But in many places, we are not sure whether the sea is actually *rising* at this unusually high rate, or the land is actually *subsiding* more rapidly than elsewhere. The relative effect would be the same. Realistically, the highest rates probably are the results of both motions combined.

The Chesapeake Bay is one of the places being monitored. What do you suppose is happening there? The bay region is undergoing one of the most rapid rises in relative sea level of any place on Earth. You can probably predict where the highest rate of relative rise in the lower bay region would be—somewhere near the crater rim. Yes, one of the greatest rates of relative sea-level rise in the Chesapeake Bay region has been measured at the Hampton Roads monitoring station (fig. 46). This station is in the lower part of the James River, near its sharp bend at the crater rim. Wouldn't Cede Cederstrom be surprised to learn of the enormous contemporary influences derived from the buried features he envisioned in another geological context so many years ago!

By now, you may have wondered about the formation of Chesapeake Bay itself. Is this grand estuary a direct result of the meteorite impact? In short, no, but the impact did help determine the bay's location. We are confident that the bay is not 35 million years old. Remember that when the meteorite struck, the target site lay under approximately 600 feet of ocean water. Chesapeake Bay

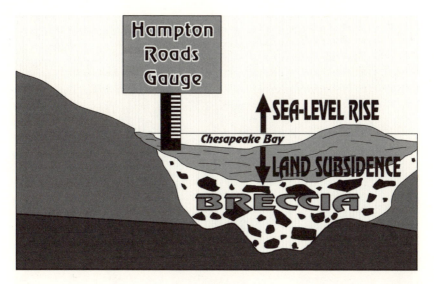

46. A conceptual diagram of relative sea-level rise near the crater rim (a combination of actual sea-level rise and land subsidence). Part of the relative rise is attributable to subsidence of the Exmore breccia and overlying sediments upon which the Hampton Roads gauge is built (at the mouth of the James River).

wasn't formed until about 10,000 years ago. But to completely understand its geological history, we need to go back a little farther to about 18,000 years ago (fig. 47).

This was a time when the great northern ice sheets were at maximum extent and thickness—the culmination of the Pleistocene ice age. Ice stretched from northern New Jersey to northern Canada, and spanned North America from coast to coast. The ice thickness approached a mile or more. Northern Europe and Asia also were buried under a similarly thick ice sheet . This vast volume of frozen water was derived from the evaporation of ocean water, which reduced sea level to around six hundred feet below present level. The sea-level drop exposed and dried out the area that is now Chesapeake Bay. Dry land extended at least fifty miles east of the present Virginia shoreline. The ancient Susquehanna and Potomac Rivers, and the rivers of Virginia, had to cut valleys across this wide coastal plain to reach the Atlantic. Curiously, these river valleys converged in the region that would eventually become the

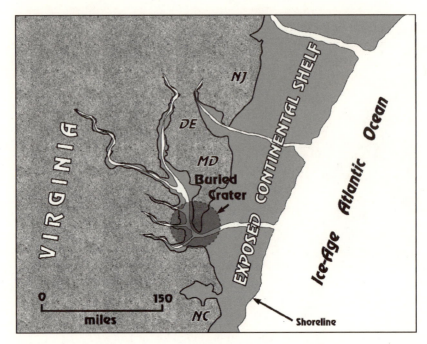

47. Paleogeography of southeastern Virginia during the last maximum buildup of continental ice sheets on North America, approximately 18,000 years ago. The circular depression overlying the Chesapeake Bay impact crater caused the ancient Susquehanna, Rappahannock, York, and James Rivers to converge there.

bay. But the convergence was not merely coincidence. There is a good geological reason for it, related to the crater's presence.

Because gravity causes rivers to seek the lowest elevations available, they flowed into the bowl-shaped depression caused by the subsiding Exmore breccia. In this way, the sagging crater floor acted as a template, which predetermined the location of Chesapeake Bay. But the bay didn't actually exist until the northern ice sheets melted ten thousand years ago, and the consequent rise of sea level flooded the convergent valleys.

CHAPTER TWELVE

Threatened Ground Waters

THE PRESENCE of the Chesapeake Bay crater also helps to explain a long-standing question about ground-water quality in southeastern Virginia. It had puzzled geohydrologists, including Cede Cederstrom, for more than fifty years. Why did so many shallow water wells, especially on the western side of Chesapeake Bay, produce unusually salty water? The water in some wells is in fact *brine*, seawater that is one and a half times saltier than normal. On a map, the distribution of brine-producing wells forms a semicircular pattern around the western side of the lower bay. Superimposed on our new crater map, the brine distribution pattern nearly matches the size and location of the crater's western sector. The saltiest ground water encountered so far comes from the Kiptopeke core hole, the site nearest the crater center. Water-quality tests show that the brine saturates the porous Exmore breccia, but is absent from shallower aquifers lying above the breccia. Recall that the impact blast completely removed all the original ground-water aquifers from a 200-square-mile region, and the resulting excavation subsequently was filled with the Exmore breccia. The breccia now serves as a single enormous brine reservoir. The brine is sealed off from younger freshwater reservoirs by the Chickahominy Formation, a 300-ft-thick confining unit of impervious clay, which overlies the breccia (fig. 48).

This brine reservoir poses a potential environmental hazard for the lower Chesapeake Bay region. The reason, as I will discuss in more detail in chapter 13, is that the Chickahominy Formation is cut by many faults, which lead down to the brine-filled breccia. The faults are defects in the geological framework; defects that might allow brine to seep upward and defile overlying freshwater aquifers.

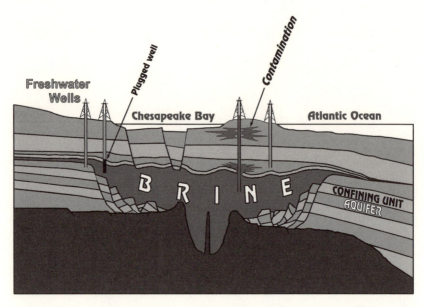

48. An interpretive cross section shows the distribution of hydrogeologic units (aquifers and confining units) within and around the Chesapeake Bay impact crater. The presence of the huge brine reservoir increases the risk of contaminating overlying freshwater aquifers, and limits the availability of deeper freshwater aquifers in southeastern Virginia.

Water wells inadvertently drilled into the brine also risk contaminating the shallow freshwater aquifers. Salty water could leak upward, for example, through spaces between the well bore and the outer surface of the drill pipe, or from rusted joints between sections of drill pipe.

The depth of burial and thickness of the brine reservoir also limit the possibility of finding new freshwater aquifers in the bay region, especially on the Delmarva Peninsula. It is likely that increased future populations in Delmarva communities will create a demand for additional freshwater supplies. Suppose the shallow freshwater aquifers were to become depleted or contaminated. Where would residents look for new sources? In areas outside the crater, new aquifers might be reached by drilling into deeper sedimentary layers. But that won't work on Delmarva. Freshwater aquifers are unlikely to be found there below the Chickahominy Formation, because, in a sense, the whole Eastern Shore of Virginia floats on the brine-saturated breccia. Below the breccia is the fractured crystalline basement, which is highly unlikely to contain fresh ground water.

Of course, the brine can be desalinized to drinking standards, but desalination is quite expensive. Clearly, a thorough understanding of the thickness, geographic extent, and depth to the upper surface of this brine reservoir is crucial to prudent management of ground-water resources in southeastern Virginia.

The presence of the Chesapeake Bay crater also helps to explain how the brine saturated the breccia in the first place. As late as 1993, before we published the meteorite hypothesis, geohydrologists in the USGS referred to a "wedge" of salty ground water beneath the western margin of the bay, particularly between the York and James Rivers. The wedge seemed to thicken and deepen toward the Atlantic Ocean. It thinned and rose toward ground level in a westward direction. Opinions varied, however, as to what caused the high salinities. Some investigators reasoned that the brine probably migrated more than a hundred miles from some of the oldest rocks beneath the Atlantic Ocean.

In this hypothesis, the journey would have begun nearly 200 million years ago in the Jurassic Period. At this time, there was no Atlantic Ocean as we know it. Its predecessor was only a narrow, shallow seaway in the center of Pangaea. Pangaea, you recall, comprised all the continents welded together into a single landmass. One end of this proto-Atlantic seaway connected to the main Jurassic ocean, known as *Panthalassa* (fig. 49). The middle of Pangaea was thousands of miles from Panthalassa, however. As a result, the closed end of the seaway was surrounded by a mid-continent desert—a terrain parched by searing temperatures and strong winds for several million years. Extremely arid conditions in this desert evaporated the salty water of the seaway at unusually rapid rates. Evaporation was so fast that the water soon became saturated with mineral salts, and they began to precipitate onto the floor of the seaway. Precipitation continued for millions of years as Pangaea gradually broke into separate continents, and the Atlantic seaway slowly widened. Before precipitation ceased, the salt bed reached several hundred feet in thickness, and covered hundreds of square

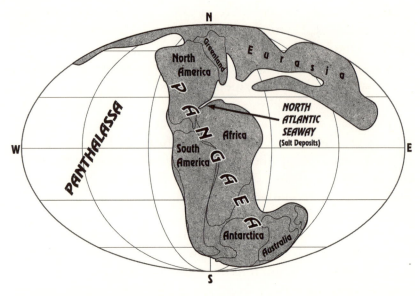

49. Reconstruction of the supercontinent Pangaea, as it appeared 200 million years ago when the Atlantic Ocean was only a narrow seaway between the North American and African tectonic plates. A single enormous ocean, Panthalassa, surrounded Pangaea.

miles. The mother salt bed is now buried two miles deep beneath the continental shelf off New Jersey (fig. 50). During the long passage of geologic time, however, slender fingers have poked up from the mother bed toward the seafloor to form salt domes. Salt domes make excellent traps for petroleum, and oil companies exploring off New Jersey have drilled the salt in a few wells.

The long-distance migration hypothesis proposes that deep fresh ground waters encountered these ancient salt beds and leached out some of the salt. Over nearly 175 million years, the salty ground water would have migrated two hundred miles to Virginia and seeped thousands of feet upward into younger beds. This hypothesis, however, does not explain why the shallow brine reservoir is limited to southeastern Virginia. The most direct westward migration route would have taken the brine into the subsurface of New Jersey. But no wells have encountered shallow brine reservoirs in New Jersey, or in the intervening states of Delaware and Maryland.

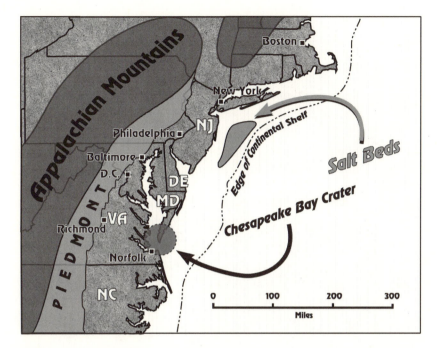

50. Ancient salt beds (200 million years old) buried deep beneath the New Jersey Continental Shelf have been thought by some investigators to be the source of the brine in the Exmore breccia.

To other USGS geohydrologists, it seems more likely that the unusually saline water in southeastern Virginia was caused by short-distance migration of salty or brackish waters directly from Chesapeake Bay or the Atlantic Ocean. This explanation is not very satisfying either, because the ground water is saltier than either of the proposed sources. Like the long-distance migration hypothesis, this one also fails to explain why the brine is limited to southeastern Virginia.

A third hypothesis proposed for the origin of the brine requires a chemical process called *reverse osmosis* (fig. 51). Many of you learned about osmosis in high school chemistry class. As you will recall, it involves two different reservoirs containing solutions of unequal salt concentration, which are separated by a finely porous, osmotic membrane. In nature, osmotic pressure causes water molecules to flow through the membrane from the low-salinity

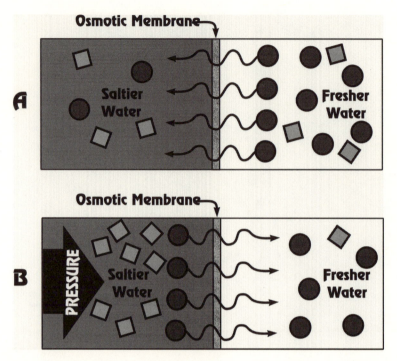

51. *(A)* In the process of osmosis, water molecules flow naturally from the fresher reservoir into the saltier reservoir until each reservoir contains equal ratios of water molecules *(spheres)* to salt molecules *(cubes)*. *(B)* For reverse osmosis to work, some source of pressure is needed to push the water molecules from the saltier reservoir into the fresher one. Salt molecules do not pass through the osmotic membrane, so the ratio of salt molecules to water molecules increases further in the saltier reservoir and produces brine.

reservoir into the high-salinity solution. The salt (sodium and chloride ions), however, does not pass through the membrane. Water continues to flow through the membrane until the ratio of salt molecules to water molecules is the same in each reservoir.

Under most natural conditions, osmosis will not work in reverse. But if some mechanism were to apply pressure to the salty reservoir, it would force water molecules into the fresh water. The consequent loss of water would raise salinity in the salty reservoir. Reverse osmosis works well in commercial desalination plants, and is used extensively to produce freshwater from seawater. Theoretically, reverse osmosis also can take place naturally in sedimentary rocks, driven by one of three possible natural mechanisms.

The first theoretical natural mechanism is rapid deposition of fine-grained sediment (clay) on top of an aquifer. This would build up a "load" pressure on the aquifer, and literally squeeze water up through the clay layer, which would act as the osmotic membrane. The second mechanism is for the aquifer to be caught up in some large-scale tectonic movement, such as the folding and squeezing that occur during mountain building. Third, is the creation of high reservoir pressures by raising the temperature of the aquifer, such as by volcanic heating. Producing a large volume of brine by any of these natural mechanisms of reverse osmosis is, however, largely hypothetical. No good examples of them have been substantiated by field studies. Moreover, to produce the Exmore brine, the tremendous heat or pressure needed to squeeze out the water would have to have been applied *after* the reservoir was sealed by the Chickahominy Formation, which would act as the osmotic membrane. We know of no post-impact tectonism or other plausible mechanism in the Virginia Coastal Plain that could have done the squeezing.

Logically, there must be another way to produce the brine. Presence of the impact crater suggests an alternative method. The meteorite strike might have produced the high-salinity brine by flash evaporation of enormous volumes of seawater over the impact site. Near the center of the explosion, the seawater and its salt would have been vaporized. But toward the margins of the blast, away from maximum temperatures, the water would evaporate and leave behind the salt. The salt would mix with breccia fragments and normal seawater as they filled the crater excavation. In addition, basement rocks beneath the brine reservoir, superheated by the impact, may have maintained elevated temperatures in the breccia for a million years. This might prolong brine production by increasing vertical flow rates of water molecules through the Chickahominy clay.

None of these hypotheses has been proven to date, but we have made rapid progress in mapping the brine reservoir. Data from seismic surveying and core drilling have allowed me to estimate the thickness, lateral distribution, and depth to the upper surface of the Exmore breccia. We no longer think of a seaward thickening wedge of salty ground water; the reservoir is more lens-shaped, and is mainly confined within the rim of the crater. The elevated salinity also extends into the surrounding ejecta blanket, however, as confirmed by ground-water tests in the Newport News core hole.

Realizing that the brine originated from the impact should further enhance our ability to predict its distribution pattern in areas lacking drill holes or seismic surveys. Our maps, cross sections, and interpretations will be available to land managers and environmental planners, who need it for enlightened stewardship of the Chesapeake's natural resources and environment.

CHAPTER THIRTEEN

Faulty Floor

IN ADDITION to the brine-filled breccia, the buried crater has produced another unexpected, potentially dangerous, geological hazard in the lower Chesapeake region. It is a system of shallow faults,
which are distributed over the crater like fractures in the wind-
shield of a wrecked automobile. The faults originate at the top of
the Exmore breccia and penetrate all the post-impact sedimentary
beds above (fig. 52). Prior to discovery of these features, geologists
believed that shallow faults were rare in the Atlantic Coastal Plain.
Few examples had been documented, and those few were not considered to be environmental hazards. However, we cannot be quite
so blasé about these crater-generated faults.

The faults linked with the crater are yet another result of the
differential compaction and subsidence of the Exmore breccia.
The subsidence apparently was not uniform across the entire breccia deposit. In some areas, seismic profiles show that mile-long
megablocks constitute part of the breccia. These blocks are relatively consolidated, and would not compact and subside as rapidly
as adjacent, less consolidated parts of the breccia. Eventually, the
upper surface of the breccia would warp, wrinkle, and produce a
wide variety of cracks, bumps, ridges, and troughs. The overlying
beds, in turn, would bend and break into many separate blocks.
The fracture planes would become faults as the blocks shifted positions under continued sediment loading.

On the seismic profiles, many of the faults extend, like inverted
root systems, for hundreds of feet upward from the top of the
breccia. Some reach all the way to, or nearly to, the bay floor and
into the river bottoms. So far, I have identified more than five
hundred locations where the compaction faults intersect seismic
profiles (fig. 53). By extrapolation, I conclude that similar faults

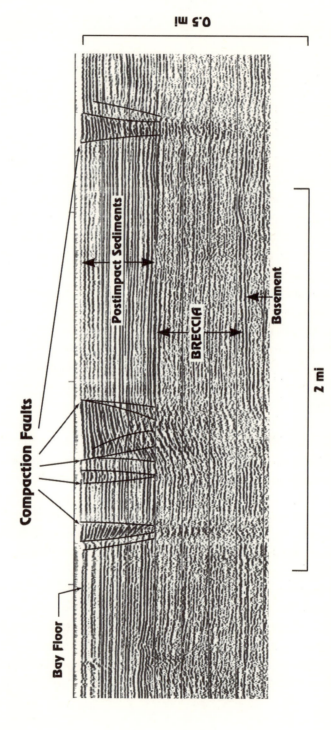

52. Expression of postimpact compaction faults on a seismic reflection profile across the Chesapeake Bay impact crater. The faults extend from the top of the breccia to the bay floor, and disrupt all the postimpact sedimentary beds.

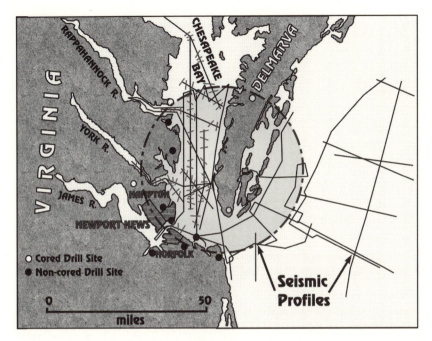

53. More than five hundred compaction faults *(short straight lines)* have been identified on the "spider web" of seismic reflection profiles (more than 1,200 miles of tracklines) that covers the Chesapeake Bay crater.

are widely distributed under all land surfaces that overlie the crater. The faults represent zones of structural weakness in the crust of the Earth. Seismic evidence indicates that fault movement has taken place in these zones from time to time for at least 35 million years. We can expect the fault blocks to continue to move, ever so slowly, for the foreseeable future, until such time that breccia compaction terminates. Most of the fault movement will be so slow that citizens of the bay region will not notice it. It will, however, continue to accelerate the long-term subsidence of coastal regions over and near the crater.

The lower bay region includes the urbanized areas of the Middle Neck and York-James Peninsulas, the heavily populated swath between Virginia Beach and Newport News, and the lower part of the Delmarva Peninsula. The lower Chesapeake Bay urban corridor is jammed with major highways, tunnels, bridges, military bases, weapons depots, airports, research facilities, and industrial plants. All are susceptible to ground instability due to continued slow movement along the shallow compaction faults.

Suppose, moreover, that a large earthquake should jolt southeastern Virginia. That possibility poses the danger of sudden large movements along some of the faults, which could threaten lives and structures around the bay. Buildings and bridges inside and near the crater rim are potentially much more susceptible to earthquake damage than those outside the rim. If the Hampton Roads Planning District Commission were contemplating construction of a new hospital complex, for example, knowledge of the fault distribution would help them decide where and how to construct it for minimum earthquake damage.

Among the earthquakes recorded in Virginia, only four, of low magnitude, originated in the Chesapeake Bay region. Curiously, all four were either inside or near the crater rim. Rest assured, however, that the probability of a major earthquake in this region is not high. Historically, the entire state has been seismically quiet. During the past two hundred years, only four quakes reached magnitude 4 on the Richter scale (moderately strong, with negligible damage; energy equivalent to about 10 tons of TNT). Nevertheless, there are many recorded examples in which a strong earthquake has shocked a historically quake-free region. So, we should not ignore the future possibility of a major jolt, as it might take only a single strong shock to heavily damage this fault-riddled region.

A detailed study of the locations, orientations, and motion histories of these compaction faults is necessary to provide citizens of the lower bay region with maximum possible protection from these geologic hazards. A principal goal of the USGS crater research has been to provide those data. But to do so we needed more seismic reflection profiles. The profiles donated by Texaco were so widely spaced apart that we could initially produce only highly generalized maps of the faults and of the brine reservoir. The need for additional detail led me to undertake a new seismic survey of the crater. In April of 1996, we contracted the research vessel *Seaward Explorer* to do the job.

CHAPTER FOURTEEN

Seaward Explorer

APRIL does not bring spring warmth to Woods Hole. Sunday, April 21, 1996, was typically cold, overcast, and foggy. The 100-ft-long *Seaward Explorer* (fig. 54) was tied up at the Woods Hole Oceanographic Institution's dock, crammed with navigation gear and seismic surveying instruments. Her low-slung work deck was crowded with a large air compressor, a six-foot-tall hydrophone reel, a heavy-duty lifting crane, a winch for the air-gun cable, several large buoys, coils of rope, chain, and spare cable, and two laboratory vans the size of house trailers. *Explorer* was off to Chesapeake Bay for a new survey of the crater. Only this time, we knew what we were searching for. Our research strategy would provide new details of the crater's structure, help pinpoint the distribution limits of the brine reservoir, and determine the spatial orientations of the compaction faults. First, we would collect a closely spaced grid of intersecting seismic profiles inside the bay. The old USGS profiles and the Texaco profiles were spread several miles apart, leaving wide gaps in the trackline grid. Second, we would collect several profiles across the continental shelf in the eastern sector of the crater, where we had only rudimentary data. Third, we would tune our instruments to gather high-resolution reflections from shallow beds, as well as from the deeper strata. Images of the shallow layers were especially important for identifying the postimpact compaction faults. The Texaco profiles were not sufficient for detailed fault assessment, because the techniques used to process the seismic signals didn't reveal reflections from the upper three hundred feet of sediments. Texaco geologists had been exploring for deeper targets.

54. The research vessel *Seaward Explorer* sets out from Woods Hole, Mass., to survey the Chesapeake Bay impact crater.

Ordinarily, the principal investigator for a seagoing research project also serves as Chief Scientist for the cruise. But in this case, health problems prevented me from going along. In my place, Debbie Hutchinson volunteered to lead the expedition (fig. 55). Debbie is a marine geophysicist, whose field studies have taken her from the U.S. Atlantic margin to the Great Lakes, the Gulf of Mexico, and even to Lake Baikal in central Russia. She also is the science manager for the USGS Woods Hole Field Center. Co-Chief Scientist was Steve Colman, an experienced Chesapeake researcher, particularly skilled in the interpretation of ice-age and modern landforms and the processes that produce them. Tommy O'Brien was our resident expert on the collection and shipboard processing of marine seismic data. His job was to make sure that the shipboard instruments produced strong seismic signals, and that returning reflections were properly transformed from acoustic to digital electronic form. David "Twig" Nichols was specifically trained to run and maintain the air gun and the hydrophone array. Barry Irwin was our satellite navigation specialist. He was responsible for pinpointing the ship's location to within a few meters' accuracy. Without his skills, we could not precisely plot the

55. The USGS scientific party that carried out the April 1996 seismic reflection survey of the Chesapeake Bay crater. *Front row, left to right:* Nancy Soderberg, Wylie Poag, Tommy O'Brien, Debbie Hutchinson; *back row:* John Evans, Jeremy Loss, Barry Irwin, Dave Nichols, Steve Colman.

location of any of the crater's features. Fortunately, we could rely on *global positioning system* (GPS) coordinates, provided by navigation satellites recently put into Earth orbit by the U.S. Navy. This system provides phenomenal navigation accuracy (to within a few feet) for seagoing vessels at any point on the world ocean.

Nancy Soderberg was our data curator. She had to keep track of all data collected and assure their safe return to the Woods Hole data center. Two computer specialists, John Evans and Jeremy Loss, were onboard to integrate the navigation data with the seismic data. John and Jeremy had to record the ship's exact location every ten seconds, precisely when the air gun fired.

If all equipment functions properly, there are no breaks in the work schedule during a seismic reflection survey. It is a 24-hour-a-day procedure. Half the scientific party must be awake and alert at all times. So Debbie divided the scientific party into two four-person watches. As long as the weather was good and no emergencies arose, one watch worked while the other slept.

The remainder of the shipboard party consisted of the ship's crew. Commanded by Captain Keith Davis, it included First Mate Kevin Rager, Engineer David Evensen, Seaman Peter O'Brien, and Cook James Bivona. The *Explorer* was a contract vessel out of Miami, Florida. The USGS and the National Geographic Society collaborated to charter her services for the duration of the crater survey.

Planning a research cruise is a fascinating but sometimes frustrating exercise. For the Chesapeake survey, we had begun to prepare more than a year earlier, when I submitted a research proposal to the National Geographic Society. I requested funding for three days of shiptime plus logistical support for dockside mobilization and demobilization. It costs $3,000-$4,000 per day to lease a 100-foot research vessel. At the same time, a parallel proposal went to the USGS for four additional days of shiptime and salary support for the scientific party. The total cruise cost would be about $100,000. Both proposals were funded in late 1995, and Debbie, Steve, and I held the first precruise planning session on January 25, 1996.

The most important decision was to select the type of seismic imaging system that would collect the data we desired. Seismic gear comes in many configurations, depending on the specific goals of an investigation. A conventional air gun, for example, releases a large air bubble each time it fires. Each bubble produces sound waves in the water column, which reverberate as the bubble ascends toward the surface. Some of the bubble reverberations interfere with seismic reflections coming from the subsurface sedimentary beds. For this cruise, we intended to use a new type of air gun, called a GI gun (Generator-Injector) (fig. 56). It was designed to reduce the bubble interference. "Twig" Nichols, our air-gun specialist, had never used a GI gun, so we sent him to Texas, where the manufacturer provided specialized training.

The GI gun required a large compressor for its air supply. The USGS did not own one, but was able to lease one from Duke University, located in Durham, North Carolina. The specific electrical input required by this compressor, however, was different from that produced by the *Seaward Explorer's* generators. This forced us to rent, in addition, a large portable electrical generator. The air

**Depth and distance
sensor**

**Air exhaust
ports**

**Air hose to shipboard
compressor**

**Air compression
chamber**

56. The GI air gun used by the USGS to provide sound waves for the seismic
survey of the Chesapeake Bay crater. The gun is approximately three feet long.

compressor was so large, however, that once it was chained to the
deck, no space remained for the generator. The only place to
squeeze it in was the roof of the laboratory van that housed the
seismic and navigation computers. This was not an ideal spot for
it. Such a heavy piece of equipment placed high above the ship's
center of gravity might cause the *Explorer* to roll awkwardly. Furthermore, we had to hope that vibrations from the compressor
would not interfere with the delicate electronic components of
the computers, and would not drive John and Jeremy nuts.

The hydrophone array was sealed inside a 300-ft-long *streamer*—
an eel-like plastic tube, three inches in diameter, which was filled
with insulating oil. The hydrophones, a series of one hundred
small acoustic receivers, would detect the seismic reflections returning from below the seafloor. The USGS had recently loaned
its streamer to colleagues at the Lamont-Doherty Earth Observa-
tory of Columbia University, however, who took it on a cruise to

Africa. So the Lamont scientists had to ship it back on short notice. The only vessel available to get the streamer back to us was notorious for frequent breakdowns, and she was scheduled to arrive in New York only a week prior to our cruise. We could only cross our fingers and hope that no delays would occur. As you readers will have noticed by now, it is almost impossible to carry out a successful scientific research or exploration program without a large dose of good luck.

Seismic surveying in some state waters requires a permit. I called Woody Hobbs at the state-supported Virginia Institute of Marine Science (VIMS), at Gloucester Point, Virginia, to inquire whether we needed one for Chesapeake Bay. Woody determined that since we were a government research group, as opposed to a commercial exploration group, a permit was not necessary. This was good news, because it avoided a possible delay while all the proper signatures were obtained.

Because the crater cruise was of special interest to the National Geographic Society, which co-funded the survey, they requested that we take photographer Ira Block onboard for two or three days to cover the trip. However, Ira didn't want to board in Woods Hole, because the first day or two would be boring, taken up with transiting to Virginia and testing the seismic gear. This meant that we had to plan for the *Explorer* to be close to a port at some point during the cruise, so that a small boat could bring Ira out to meet her. We also received requests from Woody Hobbs at VIMS and from a freelance video-news reporter, to be onboard for a day or two. Since we were planning to run a survey line up the York River, we would steam right by the VIMS campus, located on the riverbank at Gloucester Point. So we planned to pick up Woody and the video cameraman there. Finally, we invited two upper-level managers from the USGS also to board at Gloucester Point for the short York River transect.

Technical aspects of the cruise plan went smoothly under the supervision of USGS ship operations director Tom Aldrich and logistics manager Joe Newell, aided by "Twig" Nichols and Tommy O'Brien. Accommodating the visitors was to be more complicated, however. It is difficult to be at a specific location at a prearranged time when carrying out a research cruise, because many of the shipboard operations cannot be interrupted at random. Furthermore, the ship is always at the mercy of the weather and the ubiquitous Murphy's Law. The latter states that "anything that can go wrong, will go wrong." Some part of a ship's mechanical or electrical systems, or the research instruments or computers, inevitably fails at sea. Backups and replacement parts generally are on hand, but delays can hardly be avoided. So the times and places designated for taking on visitors always had to be tentative.

As logistical planning proceeded, I was busy laying out the tracklines on detailed bathymetric maps, which showed the topography of the seafloor. Much of the survey area was in shallow depths near the shoreline. Because the *Seaward Explorer* had a draft of sixteen feet, the water needed to be at least twenty feet deep along all tracklines. Furthermore, in water this shallow, the towed instruments might occasionally drag bottom. Again, a certain amount of luck was necessary so as not to severely damage or lose them. In truth, during the first day of surveying, a careless plot on one transect nearly sent the ship grinding into a sandbar whose crest was only twelve feet deep. But a sharp eye on the fathometer by one of the scientists prevented a calamity.

Then there were the crab pots. The shallow mudflats in Chesapeake Bay are prime areas for catching the blue crabs that make Chesapeake cuisine so famous. Crab pots by the thousands dot the shallow coastal waters throughout the bay and surround the river mouths like swarms of migrating butterflies. Each pot is a baited cage. It rests on the bay floor, tethered to a float, which bobs on the surface. A research vessel towing instruments several hundred feet behind it cannot maneuver nimbly enough to avoid entangling those tether lines and dislodging a few crab pots.

In addition, heavy ship traffic in some parts of the bay complicated straight-line navigation, which is best for seismic surveying. We wanted to avoid the busiest traffic lanes whenever possible, but this frequently required traversing shallow shoals where navigation was tricky.

Finally, as the fog lifted slightly around 8:00 A.M. that April Sunday, everything was in order aboard the *Explorer*. The lines were cast off, and Captain Davis slowly backed the vessel away from the dock as families, colleagues, and I waved good luck. Watching the *Explorer* slip away into the New England mist sent a wave of intense nostalgia washing over me. I was remembering the special camaraderie that develops within the confined space of a small research vessel. Such close quarters, along with the shared tensions of intense activity and heavy individual responsibilities, bring out the best and the worst in all participants. It is one of the best ways I've experienced to really get to know and appreciate your colleagues.

On the other hand, as the *Explorer* vanished, I could hardly contain my excitement that we would soon have vital new information about the crater. I anticipated that this cruise would resolve many of the remaining questions concerning its precise dimensions and detailed structure.

Alas, the late April weather was not kind to the *Seaward Explorer* once she exited Woods Hole harbor. She struggled southward against strong headwinds. Heaving seas brought queasy researchers to the deck rails for more than just the scenery. Because of the nasty weather, the *Explorer* took more than two full days (52 hours) to reach the first deployment station. This was twice the time I had estimated, so we were behind schedule even before the survey began. Once they reached the coast of Virginia, the crew deployed the seismic gear east of Parramore Island, a few miles outside the buried crater rim. The weather continued to deteriorate, however, and winds reached 30–40 knots. The seismic system could not collect high-quality data under such conditions, so the *Explorer* headed into the bay seeking calmer waters. Finding sea conditions no better inside the bay, she headed for the mouth of the York River, four days earlier than we had planned. We had to scrap plans to take on visitors from VIMS as we passed Gloucester Point, because we could not notify them in time.

Though I was not aboard, the two chief scientists kept me abreast of their daily progress and problems via a cellular phone. They found that after extremely tiring days (and nights) at sea, it was helpful to consult a relatively fresh mind. Sometimes, strong disagreements arise between shipboard scientists and advisors "on the beach," but Debbie, Steve, and I agreed on almost everything, which prevented those sorts of tensions.

At last, four and a half days after departing Woods Hole, the computers began to record good seismic data from beneath the calm waters of the York River. The *Explorer* crossed the buried crater rim a few miles upstream from Gloucester Point, approximately where we had expected it. Seven hours later, the York River survey was complete. Out in the bay, however, strong northwest winds still buffeted the surface. So Debbie decided to seek the shelter of another river—this time the Rappahannock. The weather began to improve during the Rappahannock survey, so when that transect was complete, the *Explorer* could begin its long tracklines inside the bay.

The first long line began on Thursday, April 25. The *Explorer* started a southward profile down the east margin of the bay toward the town of Cape Charles, the geographic center of the crater. Captain Davis planned to rendezvous with Ira Block, the National Geographic photographer, near the mouth of Cape Charles harbor. Ira showed up in a small skiff, right on time. After a safe transfer, the *Explorer* proceeded southward at her normal 4-knot surveying speed, toward the Chesapeake Bay bridge-tunnel. This bridge, interrupted by two tunnels so that ships can cross it, connects the southern tip of the Delmarva Peninsula to the Virginia mainland at Norfolk. At about 9:00 P.M., the lights of the northern end of the bridge loomed ahead in the darkness. At the same time, the first mate spotted a large crane-bearing barge ahead, and swung wide to the west to avoid a collision. It was a bad move. The *Explorer* was immediately swept up in a strong current and pushed onto Lattimer Shoal, a shallow sandbar. Her keel bit deeply into the top of the bar and brought the vessel to a jarring stop. Regrettably, the towed seismic instruments did not stop. Their momentum swung them up under the stern, where they caught in the propellers before the ship's engines could be shut down. Ira Block got some exciting photos in the turmoil, but the grounding was costly. The *Explorer* herself was undamaged, but the spinning propellers chewed up the hydrophone streamer. Luckily, we had a spare streamer, and the GI gun was undamaged, but the incident put us even farther behind schedule. It took two hours to free the gear, back the ship off the bar, and redeploy the instruments.

With clearing weather, our luck improved for a while. The *Seaward Explorer* steamed up and down the bay for two uninterrupted days recording excellent seismic data. Then she entered the James River to complete the survey of the crater's western sector.

On Sunday, April 28, the Atlantic at last was calm enough to allow surveying on the continental shelf. The *Explorer* exited the bay. Though the weather now befriended her, she still had to contend with what one of my professors used to call "the cussedness of inanimate objects." As soon as the ship entered the shallow Atlantic waters, the digital signal recorder broke down. The shipboard party would have to record all the continental-shelf seismic data in the old-fashioned analog format. This was not catastrophic, because Tommy and Jeremy could convert the analog signals to digital ones back in the Woods Hole laboratory. But it was one more glitch in a long string of frustrations for the seagoing scientists (and for me). It meant that additional months would have to be allotted to convert the data, and that would delay further digital processing. Analysis and interpretation, of course, also would lag behind schedule.

On the afternoon of Monday, April 29, a final bit of bad luck terminated the cruise. The *Explorer* was operating east of the bay mouth, preparing to turn north for one final long transect across the eastern half of the crater. Suddenly the hydrophones ceased transmitting signals to the shipboard computer. "Twig" and Tommy scrambled to pull the streamer aboard, and found deep abrasions in the plastic sleeve. Some elevated object, or the shallow sea bottom itself, had ground a hole entirely through the tube wall, and the invading seawater had shorted out the electrical circuits. It was difficult to estimate the extent of the damage, so Debbie could not predict how long repairs would take. Because increasingly heavy swells endangered anyone working on the wave-washed deck, she decided to call it quits and head back to Woods Hole.

Though beset by natural hazards, mechanical and electronic breakdowns, and human error, this cruise was not abnormally jinxed. Marine researchers have to withstand similar adversities on almost every cruise. The success of research onboard a seagoing vessel depends heavily on the ability of the ship's crew and the scientific party to overcome setbacks. No cruise is entirely free of gremlins. Happily, the *Seaward Explorer* arrived safely back home, having surveyed 525 miles inside and outside the crater. She brought thirty new tracklines, including 23 new crossings of the outer rim and 9 new crossings of the peak ring. Initial analysis of preliminary monitor records printed out on the ship indicated that we had an excellent new data set.

Jeremy Loss and Michael Taylor in Woods Hole, and Myung Lee in Denver, quickly set about processing the seismic data and correlating them with the navigation data. They produced beautiful profiles, which greatly enhanced my ability to interpret the geological features of the crater. The resultant improvements in structural detail and stratigraphic framework are tributes to the dedication and skill of all participants. We now could confidently document that the crater is a roughly symmetrical, subcircular excavation, approximately fifty miles wide. Its center is near the town of Cape Charles, as earlier analyses had indicated. Now that we had profiles spaced closer together, however, it was clear that the outer rim is much more irregular than the smooth curve shown on our earlier maps. Also, the eastern rim is several miles closer to Delmarva than I had originally projected.

There also was exciting new evidence that a small central peak rises three thousand feet above the center of the deep inner basin, under the mouth of Cape Charles harbor. This was not an entirely unanticipated discovery, because many complex craters have central peaks. The Texaco profiles had given no hint of a central peak, however, because until the *Explorer's* survey, no profile had crossed close enough to the center of the crater. This discovery was just the kind of result that the USGS and National Geographic Society had hoped to achieve when they agreed to fund the new seismic survey.

CHAPTER FIFTEEN

Buried Treasure

UP to this point I probably have given you the impression that the chief consequences of the Chesapeake Bay meteorite impact have a real or potentially negative effect on citizens of the bay region. But this is not necessarily the case; there also are potential benefits. One possible benefit is the likelihood that the impact produced immense economic resources (fig. 57).

More than 35 of the 150 impact craters known on Earth today contain mineral-enriched deposits (Table 1). The deposits range from common building materials, such as limestone, to oil and gas reserves, and strategic metals, like uranium, gold, copper, nickel, and platinum. The economic potential of some impacts comes in totally unexpected forms. Perhaps the best example of the latter is the 60-mile-wide Manicouagan crater in Canada. The Manicouagan crater is exposed at the surface, and its annular trough is filled by a huge, doughnut-shaped lake. The Manicouagan reservoir produces 4,000 gigawatt-hours of hydroelectric power annually. That is roughly 100 times more energy per year than was released by the Hiroshima atomic bomb, and enough to power half a million homes.

The type and volume of an impact-derived mineral deposit depend partly on the size, composition, and velocity of the meteorite. Remember, however, that for large meteorites, the stupendous impact shock completely, or nearly completely, vaporizes the impactor. Thus any large mineral deposit must come from the target rocks. That leaves two other main attributes that determine the economic potential of impact-generated mineral deposits. First is the composition of the target rocks. Are they igneous or sedimentary? What kind and size of mineral deposits were present before the meteorite struck? Second is the potential for preserving the

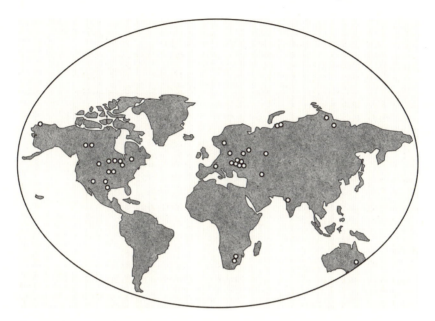

57. Distribution of 35 impact craters known to contain economic deposits.

crater through long periods of geologic time. Will the crater with-stand erosion, mountain building, or other enormous distur-bances brought about by shifting tectonic plates without being deeply buried or destroyed?

Economic deposits of impact craters can be produced in three principal ways. First, the impact might merely redistribute miner-als already present in the target rocks. Deeply buried ore veins, for example, might be brought to the surface by the impact. Sec-ond, the impact generally melts a portion of the target rocks into thick sheet-like bodies. The melting process concentrates or en-riches mineral deposits already present. Third, mineral deposits might form in an impact crater millions of years after the meteor-ite struck. For example, oil and gas could migrate into a buried breccia reservoir and be trapped there by impervious overlying rock layers. Whatever the mechanism of formation, the total eco-nomic value of known impact-derived mineral deposits is not triv-ial. In 1994, the total gross value for North America alone amounted to $5–6 billion.

TABLE 1
Known Economic Deposits Found in Terrestrial Impact Craters

Crater	Diameter (miles)	Economic Deposit
Ames, Oklahoma	9.5	Oil & gas
Avak, Alaska	7.2	Gas
Barringer (Meteor), Arizona	0.7	Silica
Beyenchime-Salaatin, Russia	4.8	Pyrite
Boltysh, Ukraine	14.4	Oil shale
Carswell, Canada	23.4	Uranium
Charlevoix, Canada	32.4	Ilmenite
Crooked Creek, Missouri	4.2	Lead, zinc
Decaturville, Missouri	3.6	Lead, zinc
Ilyinets, Ukraine	2.7	Agate
Kaluga, Russia	9	Mineral water
Kara, Russia	39	Diamond, zinc
Logoisk, Belarus	10.2	Amber, calcium phosphate
Lonar, India	1.1	Various salts
Manicouagan, Canada	60	Hydroelectricity
Marquez, Texas	13.2	Oil & gas
Obolon, Ukraine	9	Oil shale
Popigai, Russia	50	Diamond
Puchezh-Katunki, Russia	48	Diamond, zeolite
Ragozinka, Russia	5.4	Diatomite
Red Wing Creek, N. Dakota	5.4	Oil & gas
Ries, Germany	14.4	Lignite, bentonite, moldavites
Rotmistrovka, Ukraine	1.6	Oil shale
Saint Martin, Canada	24	Gypsum, anhydrite
Saltpan, South Africa	0.7	Various salts
Serpent Mound, Ohio	4.8	Lead, zinc
Siljan, Sweden	33	Lead, zinc
Slate Islands, Canada	18	Gold
Steen River, Canada	15	Oil
Sudbury, Canada	150	Copper, nickel, platinum
Ternovka, Ukraine	7.2	Iron, uranium
Tookoonooka, Australia	33	Oil
Ust'-Kara, Russia	15	Diamond
Vredefort, South Africa	180	Gold, uranium
Zapadnaya, Ukraine	2.4	Diamond
Zhamanshin, Kazakhstan	8.1	Bauxite, impact glass

One of the recent surprises in the exploration for oil and gas has been the realization that large quantities of petroleum are trapped in ancient impact structures. Buried impact craters in North Dakota, Alaska, Oklahoma, Canada, and Australia are notable for their petroleum accumulation. The most prolific oil-producing crater is the Ames impact structure in Oklahoma. But not until 1992 did geologists recognize the true origin of the structure. Fifty-two wells currently produce from the Ames crater. They are expected to yield 50 million barrels of oil and 60 billion cubic feet of gas before the reservoirs are depleted.

Precious and strategic minerals are mined from many impact craters. Ukrainians have produced iron and uranium ores from Ternovka crater for more than 50 years. Total reserves of more than 700 million metric tons are estimated for this crater. The gigantic Vredefort crater in South Africa, 180 miles in diameter (more than triple the size of the Chesapeake Bay crater), produces uranium and gold. In fact, more than half the gold ever produced has come from the Vredefort mines. Production there began more than 100 years ago, and the total value of extracted gold amounts to at least $50 billion. More than $4 billion worth of uranium has been extracted along with the Vredefort gold.

The most notable example of impact-generated economic deposits, however, is located in Ontario, Canada. Ontario's Sudbury mining complex is a world-class producer of copper, nickel, and platinum ores. The ores are taken from an unusually thick (1.5 miles) impact melt sheet, which so far has yielded 1.5 billion metric tons of ore. The current production is worth $2 billion per year.

The origin of the Sudbury structure has been debated for decades. A particularly puzzling aspect has been the asymmetrical shape of the structure. Its outline resembles a kidney more than a circle. The Sudbury impact took place nearly 2 billion years ago, however, and since then has undergone periods of intense tectonic squeezing and folding, followed by extensive erosion. What is left is only an irregular eroded remnant of a crater originally nearly 150 miles wide. In the last few years, abundant diagnostic evidence of shock alteration has been documented from rocks within and around the Sudbury structure, which proves its impact origin.

Crystalline target rocks, such as those of the Sudbury and Chesapeake Bay craters, characteristically yield enormous melt sheets when struck by a large meteorite. The postulated Chesapeake Bay melt sheet has not been identified yet, but its potential volume can be estimated from studies of other craters. A maximum estimate would be approximately 2,500 cubic miles. If it could be mined, potential mineral deposits in a melt sheet this size could make an enormous contribution to the economic prosperity of the bay region.

How do we pinpoint the location of such a deposit? A melt sheet normally forms deep within the inner basin of a large impact crater. Sometimes its presence can be detected on seismic profiles, but so far I have not been able to identify a melt sheet on the Chesapeake Bay profiles. I suspect that the available profiles do not image deeply enough into the crater to reveal the sheet. The most reliable detection method is to drill deep into the center of the crater. The best location for a deep drill hole into the Chesapeake Bay crater is somewhere near the town of Cape Charles. A borehole about five thousand feet deep would be necessary to reach the expected level of the melt sheet. The estimated cost of such a borehole would be roughly half a million dollars.

CHAPTER SIXTEEN

Subterranean Waste

GARBAGE DUMPS receive special attention from today's "advanced" civilizations. We don't have enough of these sites, and many contain hazardous contaminants. Industrialized nations are scrambling to find safe, socially and politically acceptable places in which to store wastes that could seriously harm the environment and us. We all know the potential long-lasting danger from radioactive components left over from nuclear weapons development. Elaborate and very expensive schemes have been proposed to store and protect these highly dangerous substances, but all current storage facilities are temporary.

The most highly publicized long-term nuclear storage facility is the proposed site inside Yucca Mountain, about one hundred miles northwest of Las Vegas, Nevada. The site has undergone intense geological and hydrological analysis for about twenty years, at a cost of billions of dollars. This has been merely an initial study to determine whether the facility could prevent environmental contamination for ten thousand years. Billions more dollars and years of work would be required to actually construct and fill the repository.

Many so-called low-level wastes, however, such as paints, motor oils, dry-cleaning chemicals, and pesticides, require less demanding disposal efforts. Unfortunately, we have been too careless with these items in the past, as the news media point out every day. My fellow citizens of Cape Cod and I, for example, are reeling from recent revelations about massive chemical contamination of our sole ground water aquifer. Leaking jet fuel, chemicals from exploded weapons, and huge buried dump sites on a nearby military base have created ten or more contaminant plumes within the Cape's fresh ground-water system. The southward dip of the buried aquifer is leading the plumes directly toward Falmouth, the second largest town on the Cape.

It is quite clear that we need to find new ways to safely get rid of such wastes. Disposing of them, however, is no simple matter. There are many regulatory as well as technical complications. For example, since about 1970, the U.S. federal government has produced regulation after regulation to govern the assessment, monitoring, and remediation of hazardous wastes in the subsurface. To name a few, there are the National Environmental Policy Act (NEPA), the Clean Water Act (CWA), the Safe Drinking Water Act (SDWA), the Toxic Substance Control Act (TSCA), the Resource Conservation and Recovery Act (RCRA), and the Comprehensive Environmental Response, Compensation, and Liability Act (CERCLA). On top of these, individual states have their own sets of regulations.

On the technical side, methods for injecting liquid industrial wastes into subsurface rock formations have been developed and refined for nearly 70 years, since first introduced in the 1930s. As of 1985, 90 facilities in the U.S. were injecting subsurface wastes through 195 wells. Most of these facilities are used by chemical, petrochemical, and pharmaceutical companies to dispose of more than 300 different varieties of wastes.

What does this have to do with impact craters? It turns out that the most common subsurface geologic setting into which hazardous wastes are injected is deep saltwater aquifers. Remember that the Exmore reservoir is just such an aquifer. The water is too salty for drinking and for most industrial and agricultural uses, and desalination would be unusually expensive because of the high concentration of salts. Thus, if we have no other use for the brine, we might inject low-level wastes into it without significantly degrading the environment (fig. 58).

The latest extensive evaluation of potential waste-storage sites beneath the Atlantic Coastal Plain was carried out nearly 30 years ago, long before I identified the Chesapeake Bay impact crater. USGS geohydrologists Phil Brown and Marjorie Reid, who did the evaluation, focused on saline reservoirs, including those beneath and around Chesapeake Bay. Brown and Reid concluded that, as

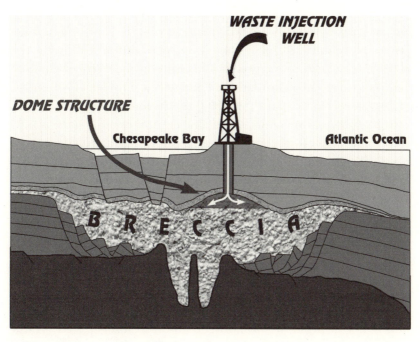

58. Interpretive cross section of the Chesapeake Bay impact crater showing a potential method for storing wastes within the brine-filled Exmore breccia.

a rule of thumb, the best places for storage would be brine-filled sand layers more than 20 feet thick, sandwiched between confining clay layers at least 20 feet thick, especially if they were buried 1,500 feet or more below sea level. After having examined selected wells in Virginia, these researchers concluded that, in the Chesapeake region, sand beds in the lower part of the Potomac Formation have the greatest potential for safe storage of low-level wastes. They constructed maps to show the depths to the tops of the sands, the sand thicknesses, and their lateral distributions.

I'll bet some of you can predict what the maps showed for southeastern Virginia. Brown and Reid projected the potential waste-storage sands right across lower Chesapeake Bay without interruption. We know now, however, that these sands are not there. They were all vaporized or excavated by the late Eocene meteorite impact. In their place is the brine-filled Exmore breccia, which has a different distribution, thickness, and depth to its upper surface than the deep sands of the Potomac Formation. The breccia does, however, fit Brown and Reid's general criteria for good storage potential. It is at least half a mile thick. It is confined above by the 300-ft-thick Chickahominy clay, and below by the dense crystalline basement, which is thousands of feet thick. Furthermore, much of the breccia is deeper than 1,500 feet below sea level.

Remember, however, that the Chickahominy Formation is fractured and faulted in many places, which might provide escape routes for stored wastes. We would need to find areas in which the upper surface of the breccia is arched into broad, unfaulted domes. Our seismic profiles display many of these. We could then drill into the tops of the domes, pump liquid waste into the briny reservoir, and seal the borehole. The Chickahominy would prevent upward migration of the waste, protecting the shallow freshwater aquifers. The curved walls of the domes would prevent lateral flow, provided the domes were not pumped completely full. In order to be sure the roofs of the domes are not cut by compaction faults, however, we would need an even closer-spaced seismic grid than we have at present.

Though engineers have developed the technology for deep fluid injection, it is unlikely that today's sociopolitical mind-set and regulatory morass would allow such disposal in fragile environments like Chesapeake Bay. Fifteen states regularly dispose of wastes in the subsurface, but Virginia is not among them. Scientific knowledge and engineering skills are often ahead of society's ability or willingness to use them, and this may be a case in point. The Exmore brine-storage concept might be useful for future generations, however. As population growth continues in southeastern Virginia, accompanied by expanding military, industrial, agricultural, and research facilities, this area might soon be desperate for innovative waste repositories.

CHAPTER SEVENTEEN

Chicken Little's Dilemma

CHICKEN LITTLE claimed the sky was falling. Her friends believed it and joined her journey to warn the king. They didn't get far. The trip was terminated, unfortunately, by Foxy Loxy's appetite for chicken flesh. But Chicken Little was right. Ask the young Ugandan, who was struck in the head by an inch-sized meteorite on August 14, 1992. He was lucky that the tiny missile had first hit a tree and slowed down. Even a meteorite that small would easily have killed him had it maintained its original velocity of 25 miles per second. Larger chunks of falling "sky," like the Chesapeake Bay meteorite, have played a major role in the evolution of the Earth, its living occupants, and of all the other rocky planets and moons in our solar system. A casual glance at modern satellite images of our Moon, for example, reveals hundreds of thousands of impact craters. They range in diameter from a few feet or less to thousands of miles. Large craters are pocked with smaller craters, which are pitted with even smaller craters, like fleas upon fleas upon fleas. Mercury, Mars, Venus, and most of the planetary moons also are densely cratered. Their impact-scarred terrains are clear evidence that our solar system has been undergoing heavy meteorite bombardment for billions of years.

Where did all these meteorites come from? Most are derived from clusters of asteroids and comets that orbit our Sun. Because humans tend to think of our solar system in terms of a historical timescale, it seems orderly and unchanging. But on a geological or astronomical timescale, the system continually changes, and the bodies that comprise it interact, sometimes catastrophically. Perhaps the best expression of this dynamic interaction took place in July of 1994. During an entire week, scientists worldwide were enthralled to watch twenty-one huge pieces of the fragmented

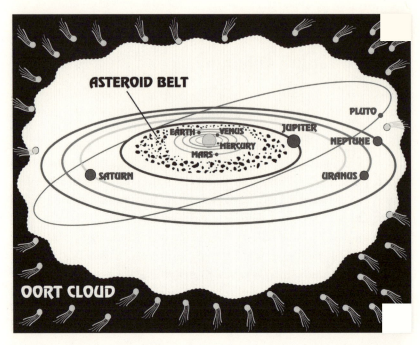

59. Most meteorites that strike Earth are believed to be either rocky or metallic asteroids from the asteroid belt that orbits between Mars and Jupiter. Other meteorites are the remains of icy comets derived mainly from the Oort cloud, which orbits beyond Pluto.

comet Shoemaker-Levy 9 bombard the giant gas planet Jupiter. Expanding gas plumes larger than the entire Earth were blasted from Jupiter's atmosphere by single fragments of the comet as they impacted the surface.

Comets are made of mainly ice combined with lesser amounts of rocks and dust. These icy bodies are concentrated in the outer edge of our solar system, far beyond the orbit of Pluto, Earth's most distant planetary neighbor. Out there, comets aggregate in a region known as the *Oort cloud* (fig. 59). Gravitational disturbances, caused by enormous clouds of interstellar dust or passing stars, can nudge comets out of the Oort cloud and send them hurtling into the inner solar system. Once there, comets have tremendous potential for impacting and violently rearranging planetary surfaces.

Comet Hale-Bopp, which crossed Earth's path in the spring of 1997, provided a stark perspective of the potential threat from such impactors. Hale-Bopp is gigantic—four times larger than the Chicxulub meteorite, whose ancient impact contributed to the massive K-T extinction event. A small shift in Hale-Bopp's next orbit past Earth could permanently redirect the future of mankind.

Asteroids are even more dangerous adversaries, because they reside closer to Earth, and cross our orbit much more frequently. Asteroids are pieces of rocky or metallic debris left over from the formation of our solar system—material that wasn't used to construct planets. Most of this debris has collected in the well-known asteroid belt, which orbits the Sun between Mars and Jupiter. Asteroids range in size from dust particles to irregular fragments hundreds of miles across. Some are occasionally knocked out of their normal orbits by the gravity attraction of planets, or by ramming into one another. When this happens, they usually break into smaller fragments, some of which end up blasting craters into nearby planets or other asteroids. Some asteroid collisions have been violent enough to propel chunks of Mars all the way to Earth, more than a hundred million miles away. One Mars fragment collected from the ice fields of Antarctica made the headlines recently, because it was thought to contain possible evidence that micro-organisms once inhabited Mars.

The most spectacular example of damage caused by colliding asteroids has been documented on the asteroid Vesta. Vesta is one of the largest asteroids identified so far, with a diameter of 315 miles. Amazingly, Vesta's south pole displays an impact crater which spans nearly the whole diameter of the asteroid. The crater is 270 miles across and 8 miles deep, with a central peak 8 miles high. More than 1 percent of the entire asteroid would have been ejected into space by this calamitous impact. Some asteroids, such as Vesta, have unusual chemical compositions, which make them easy to identify. Scientists believe that they have collected fragments of Vesta on Earth's surface.

A widely accepted impact hypothesis proposes that the Moon originated from an ancient collision between a Mars-sized asteroid and Earth. A gigantic chunk, or several chunks, of the Earth's crust may have been blasted into space to eventually become our Moon. More recent research has suggested, however, that the Earth impactor was twice the size of Mars. The huge volume of rock ejected from that collision may have initially formed a debris ring around Earth, similar to that of Saturn. Eventually the debris would have consolidated to form the Moon.

The erratic, bumper-car behavior of the nearest asteroids and comets, collectively called near-Earth objects, or NEO's, casts them as potential natural adversaries of mankind. But isn't it odd that our own planet seems to have avoided this meteoritic barrage? Well, the evidence is deceiving. Though Earth's surface is pocked with fewer than two hundred identified craters, there are undoubtedly many more to be found. You might find it difficult to imagine that hundreds of thousands, or even tens of thousands of impact structures could be lurking out there. Even so, there probably were that many strikes, most concentrated in the first few billion years of our planet's history. Earth, like her sister planets, has been in the middle of the meteorite firing range for her entire existence. The pertinent question is, where did the craters disappear to?

The answer lies in the fact that the Earth has a combination of unusually active erosional and depositional systems, plus large ocean basins and a highly mobile veneer (geologically speaking) of shifting and sinking tectonic plates. This combination of processes, unique in our solar system, has effectively erased and/or buried Earth's crater record. The discovery rate of earthly craters has accelerated recently, however, due to intensified searches aided by satellite imagery and other remote sensing instruments. The current discovery rate of Earth craters is about three to five per year.

Using the Moon's multicratered surface as a standard record, planetary scientists have divided meteorites into several categories according to the size of craters they produce. Those data are used, in turn, to estimate the rate at which meteorites from each size-category have bombarded the Earth-Moon system since its origin. The annual rain of small meteorites, most of which vaporize in the Earth's atmosphere, is continuous and heavy; around 25,000 per year. Only a tiny partial record of them has been recovered— about 10,000 individuals altogether. There also is a steady downpour of micrometeorites, averaging 0.004 inch in diameter, which adds roughly 20,000 *tons* of material per year to the Earth's surface. The ice sheets of Antarctica and Greenland are particularly prolific collectors of small meteorites and micrometeorites. As upper layers of the ice erode over thousands of years, the meteorites remain behind, concentrated like beach pebbles on the icy surface. The East Antarctic ice cap alone has yielded 5,000 small meteorites, and is believed to contain at least 760,000 specimens in total. The number of micrometeorites collected from both ice sheets is more than 100,000.

There is a steady fall of larger meteorites, as well. Recently declassified military satellite images reveal that on average, one house-sized meteorite per month explodes and disintegrates in Earth's upper atmosphere, without ever reaching the surface. But theoretically, meteorites greater than a half-mile in diameter should survive the atmospheric transit and strike the continents or oceans.

Some experts have used these cratering rates to assess the risk of a person being killed by a meteorite impact during a lifetime of 70 years. The statistical results suggest that the risk to a U.S. citizen is about 1 in 20,000, or about the same chance as being killed in an airplane accident (fig. 60). These estimates are only general guidelines, however, and must be applied with caution. For example, this assessment suggests that a meteorite of 2–3 miles diameter, like that of Chesapeake Bay, should strike Earth every 50–60 million years. Yet Popigai, which is in the same size category, struck within less than one million years of the Chesapeake Bay impact.

Given the immensity of the total documented and estimated impact record, we have little choice but to consider NEO's as serious threats to civilization as we know it. Are we, then, at the complete mercy of these random (or possibly periodic) natural events? Or is there something we can do to detect and deter NEO's before they strike our home planet?

Chances of Dying From Selected Causes (USA)

Motor Vehicle Accident	1 in 100
Murder	1 in 300
Fire	1 in 800
Firearms Accident	1 in 2,500
Meteorite (lower limit)	1 in 3,000
Electrocution	1 in 5,000
Meteorite	1 in 20,000
Passenger Aircraft Crash	1 in 20,000
Flood	1 in 30,000
Tornado	1 in 60,000
Venomous Bite or Sting	1 in 100,000
Meteorite (upper limit)	1 in 250,000
Fireworks Accident	1 in 1,000,000
Food Poisoning	1 in 3,000,000

60. David Morrison and Clark Chapman estimated that the risk of a U.S. citizen being killed by a meteorite impact during a 70-year life span is about the same as that of dying in an airplane crash: 20,000 to 1.

Scientists have proposed a number of options for dealing with the impact hazard. The key is to be able to detect one far enough in advance to deflect or destroy it far out in space. In 1990, scientists at the University of Arizona conceived the Spacewatch program to detect NEO's. They used the university's Kitt Peak telescope to discover and plot the orbits of several hundred half-mile-sized asteroids. Solid-state electronic sensors attached to the telescope detected numerous additional objects as small as twenty-five feet in diameter.

Spacewatch's most dramatic discovery so far is an asteroid named 1997 XF11. This mile-wide projectile was discovered in December 1997, and is headed toward Earth. Preliminary calculations of its future orbit indicated that XF11 might pass as close as 30,000 miles to Earth at 1:30 P.M. Eastern Daylight Time, on December 26, 2028. A more thorough study of a larger data set, however, subsequently changed the passing distance to more than 600,000 miles. A number of factors could shift its orbit before XF11 approaches Earth, but we have almost 30 years to track it, refine the orbital calculations, and prepare for a possible impact.

The U.S. Congress, in 1993, asked NASA to propose a more comprehensive "Safeguard Survey," which could detect all NEO's greater than half a mile in diameter, plot their orbits, and monitor their trajectories. In response, NASA devised a collaborative project with the U.S. Air Force to put sensors on satellite-tracking telescopes. They estimated that this system could inventory all NEO's in the half-mile category in about 10 years. The first test of the system, called Near Earth Asteroid Tracking (NEAT) was highly successful. An Air Force telescope on Mt. Haleakalā, in Maui, Hawai'i, detected more than 10,000 asteroids. Ninety-nine of them are larger than half a mile in diameter, and orbit within 5 million miles of Earth. This is the dangerous category. The estimated total in the dangerous half-mile size range is 1,500. An asteroid this large is expected to strike Earth an average of once every 150,000 years.

Chicken Little reappeared early in the twentieth century in the guise of Gene Shoemaker, affectionately known as "SuperGene," the founder of modern planetary geology. Among many stellar achievements, some of which I mentioned in chapter 9, Shoemaker led the team that discovered the famous Jupiter impactor, Shoemaker-Levy 9. No Foxy Loxy could deter Gene's inspired genius, so this time the message of "falling sky" reached the halls of political power. In fact, President George Bush personally presented the National Medal of Science to Gene Shoemaker in 1992. Gene was a leading proponent of a national program to inventory NEO's. In a recent summary statement to Congress, he pointed out that "if all the potentially threatening asteroids were discovered, . . . the risk to Earth would no longer be a matter of chance. We would know whether a collision is imminent. The time of impact could be predicted centuries in advance, and the place of impact could be predicted fairly accurately decades in advance."

What happens once we detect a NEO headed for Earth? Can it be diverted or destroyed? No one is absolutely sure, but five basic plans have been discussed. All five depend on the ability to deliver an impulse of energy to, or near enough to, the NEO to change its orbit. Plan one would divert the object with a laser, probably based on Earth, but possibly beamed from an orbiting satellite. Currently, however, no laser exists that could produce sufficient

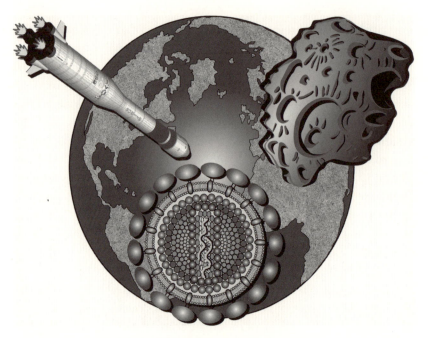

61. Three potential global catastrophes face modern civilization: nuclear war, an AIDS epidemic, and a large meteorite impact.

power to project a beam completely through Earth's atmosphere with enough energy left to deflect the NEO.

Plan two would shoot the NEO with a non-nuclear, hypervelocity projectile. In other words, hit it with a man-made meteorite. Plans three, four, and five would use nuclear devices: plan three would detonate the nuclear device on the surface of the NEO; plan four would bury the device in a shallow excavation; plan five would bury the device deep beneath the surface.

All the plans run the risk of fragmenting the NEO and producing multiple impacts. Collectively, these might be more devastating than a single large impact. Critics point out that the presence of nuclear devices powerful enough to change an asteroid's orbit might be potentially a greater danger to mankind than the threat they were designed to eliminate. Noted astronomer Carl Sagan saw a potential threat in the very ability to shift an NEO's orbit. It might tempt some group to steer an asteroid to a selected Earth target for political or economic gain.

It is abundantly clear now that Chicken Little's alarm must be taken seriously. The sky has been falling for billions of years, and will continue to do so for billions more, with no end in sight. It has been stated that today, Earth's civilizations face three potential calamities capable of wiping out most if not all of mankind: a nuclear war; a viral epidemic, such as AIDS; and a meteorite impact (fig. 61). We have come to grips with the first two and are progressing toward solutions. It is time to do likewise with the third. Understanding the nature and consequences of impact craters, such as the one beneath Chesapeake Bay, will help immensely in this task.

Glossary

agglutinated — Constructed of individual grains of sediment, as in the agglutinated test of foraminifera.

air gun — A cylindrical steel pressure-chamber towed behind a seismic survey ship. The air gun emits a bubble of compressed air at carefully timed intervals to provide sound waves that penetrate the seafloor and produce seismic reflections.

ammonite — An extinct group of tentacled, squid-like animals whose enrolled, ornamented shells are among the most abundant and diagnostic fossils of the Mesozoic Era.

angiosperm — A plant which produces true flowers, and whose seeds, enclosed in an ovary, comprise the fruit. Besides flowering trees, angiosperms include such plants as grasses, orchids, and roses. The first appearance of angiosperms was a major evolutionary event of the Cretaceous Period.

annular trough — The wide, relatively flat, circular region of a complex impact crater that separates the outer rim from the peak ring.

aquifer — A porous rock unit that contains water in fractures or in spaces between grains; the pores are interconnected, making the rock permeable, so that the water will flow when a well is drilled into the aquifer.

argillite — A fine-grained rock (claystone or siltstone) that has undergone mild metamorphism (heating or squeezing), which increases its density.

asteroid — One of a multitude of rocky or metallic bodies ranging in size from inches to hundreds of miles, most of which orbit the Sun between Mars and Jupiter.

benthic — Dwelling on the seafloor, a river bottom, or a lake bottom.

block — A rock clast the size of a house or trailer-truck, which shows little sign of wear caused by transportation or weathering.

boulder — A somewhat rounded rock clast larger than 10 inches in diameter.

brachiopods — Two-shelled benthic marine animals resembling clams, but most of which were attached to the seafloor, like mussels. Brachiopods were among the most abundant and important fossils of the Paleozoic Era.

breccia — An Italian word for a rock unit composed of irregular, angular fragments of preexisting rocks mixed together into a matrix of finer-grained particles.

brine — A water body containing dissolved salts in concentrations higher than normal seawater.

brontothere — One of a group of huge, rhinoceros-like mammals that are among the most abundant and characteristic fossils of the middle Eocene Epoch.

calcite — The primary mineral in limestone and in the tests of foraminifera and nannofossils. The calcite molecule, which contains calcium, carbon, and oxygen, is known as calcium carbonate.

calyx — The tulip-like head at the apex of a crinoid stalk. The calyx contains the animal's body and feeding apertures and flexible tendrils that help capture food from the water column.

central peak — The raised peak in the center of many, but not all, complex impact craters.

clast — A fragment or particle of rock derived from a preexisting rock body.

cobble — A rock clast larger than a pebble and smaller than a boulder, 2.5–10 inches in diameter.

coesite — A high-pressure form of silica whose density has been increased by the pressure of a meteorite impact.

confining unit — A rock unit that is impervious, or nearly impervious, to the flow of water, thereby confining water to adjacent, permeable aquifers.

core — A cylindrical section of rock. Cores allow geologists to examine subsurface rocks in three dimensions; they can tell the top from the bottom, and see the spatial relationships between different mineral grains and fossils.

Cretaceous — The period of geologic time between 140 and 65 million years ago. Dinosaurs were the dominant terrestrial organisms during the Cretaceous; they became extinct at the end of the period, coincident with the Chicxulub meteorite impact in Yucatán, northeast Mexico.

crinoid — One of a group of mostly stalked tulip-like animals, which were abundant members of Paleozoic seafloor communities.

crystalline basement — A dense, consolidated layer of igneous and metamorphic rocks that underlies the unconsolidated and semiconsolidated rocks of Earth's crust.

cyanobacteria — Green and blue-green bacteria. These organisms were the most important microbes in the Proterozoic Era, having produced the atmospheric oxygen necessary for higher life forms to evolve.

debriite — A rock unit composed of debris, which reaches its place of deposition by flowing down a slope under the force of gravity.

differential subsidence — Referring to the fact that adjacent parts of the Earth's crust subside at different rates. For example, the crust over the unconsolidated Exmore breccia subsided at a greater rate than the semiconsolidated sediments outside the crater. This differential subsidence created a topographic depression over the crater.

dinoflagellate — One of a group of plant-like planktonic microorganisms, whose abundant fossils are important for dating ancient marine rocks.

dolomite — A type of limestone in which the element magnesium is added to the calcium carbonate.

drill cuttings — Small fingernail-sized chips of rock produced by the grinding action of a drill bit.

drill string — The long string of individual drill pipes that are screwed together in order to drill a well.

drilling mud — A viscous fluid pumped down a borehole to flush out the drill cuttings and lubricate the drill bit.

ediacarans — A group of enigmatic fossils, whose flattened body impressions form a distinct community of soft-bodied organisms, which flourished just prior to the Early Cambrian appearance of hard-bodied animals.

ejecta blanket — The rock debris ejected from an impact crater, which forms a blanket-like halo outside the crater rim.

fault — A fracture in the Earth's crust along which the crust is able to move; one side moves in the opposite direction from the other side.

feather star — One of the group of modern crinoids, which are not attached to the seafloor, but are free to swim about.

feldspar — A group of abundant rock-forming minerals, including microcline, orthoclase, and plagioclase.

foraminifera — A group of hard-shelled microorganisms whose prolific fossil remains are widely used to interpret the age and depositional environments of marine rocks.

gigapascal — The unit used to measure the tremendous pressure produced by a meteorite impact.

glauconite — A green or black sedimentary mineral, which when abundant in a formation, gives rise to the term greensand.

global positioning system (GPS) — A system of navigation that relies on orbiting satellites for unusually high accuracy of location (within a few feet) almost anywhere in the world.

gneiss — A hard, coarse-grained metamorphic rock, which displays bands of different-sized and different-colored minerals.

granite — The most common igneous rock of Earth's continents.

greenhouse gas — One of a small group of atmospheric gases, such as carbon dioxide, which permit the sun's radiant energy to pass through to the Earth's surface, but which retard the escape of that energy when it is reflected back from the surface. Thus the gas acts in the same way that the glass panes of a greenhouse keep the plants within it warm.

ground zero — A term derived from weapons testing, which refers to the location where the explosion begins.

hydrogeologic unit — A rock formation that is classified either as a water-bearing aquifer or a water-retarding confining unit.

hydrophone — A small sensor towed behind a ship during a seismic survey to detect the sound waves reflecting back from subsurface rock layers.

hypercane — Winds generated by meteorite impacts, which far exceed the ~250 mph maximum velocity of a hurricane.

hypsodont — One of a group of primitive, dachshund-sized, hoofed mammals common in North America during the early and middle Eocene Epoch.

igneous rock — A rock unit that hardens from a fluid melt.

impact breccia — A specific type of fragmental rock produced by the fragmentation of existing rock formations by the enormous shock energy of a meteorite impact.

inner basin — The deepest part of a complex impact crater, located inside the peak ring.

inverted stratigraphy — Rock layers that have been turned upside down, so that the oldest layers are on the top rather than the bottom; a common condition along the raised outer rims of impact craters.

iridium — A rare metallic element that is significantly more abundant (though still in small quantities) in extraterrestrial dust and larger planetary bodies than it is on Earth. An unusual enrichment of iridium at the K-T boundary was the main clue to the Chicxulub impact, which coincided with the last of the Big Five mass extinctions 65 million years ago.

isotopes — Elements, such as oxygen, or argon, having identical numbers of protons in their atomic nuclei, but different numbers of neutrons, which gives them different atomic weights. Specialized instruments can detect the weight difference and count the number of each different isotope; this yields isotopic ratios, which can be used to interpret the age or environment at the time a rock unit was deposited.

lechatelierite — One of several high-pressure forms of silica produced by meteorite impacts.

mass extinction — A sudden disappearance (within roughly a million years) of 5% or more of the species inhabiting the Earth.

megablock — Unusually large blocks of rock, measured in thousands of feet or even miles; commonly produced by breaking up a formation with the intense shock of a meteorite impact.

metamorphic rock — A rock unit that has undergone intense heating or pressure, which alters the chemical composition and physical characteristics of the unit.

meteorite — A natural body of extraterrestrial origin, which survives passage through Earth's atmosphere and reaches the surface.

microbial mat — A paper-thin layer of bacteria and sedimentary debris. Stacked layers of microbial mats formed stromatolites, which are the principal fossil remains representing most of Proterozoic time.

microfossil — The fossilized remains of a microscopic organism.

micropaleontologist — A paleontologist who specializes in the study of microfossils.

microtektite — A microscopic bead of glass produced when a meteorite impact melts silica in the Earth's crust and sprays it into the atmosphere.

multituberculate — One of a group of small Eocene mammals with squirrel-like bodies and prehensile tails. With chisel-like incisors and multiple bumps (tubercles) on their grinding teeth, these animals were the competitors of primitive rodents in the Eocene.

nannofossil— One of a group of microscopic plant fossils, formed of calcite plates, which are widely used to date marine rocks and to determine the environments in which they were deposited.

North American tektite strewn field— A broad area of the western North Atlantic Ocean, Caribbean Sea, Gulf of Mexico (and perhaps the Southern Ocean), and adjacent land areas, in which 35-million-year-old impact ejecta has been identified; this debris is correlated with the Chesapeake Bay impact.

Oort cloud— A vast group of comets which orbit together beyond Pluto; individual comets can be displaced from the cloud and collide with the planets and satellites of our solar system.

Pangaea— The supercontinent that formed from the collision of all Earth's continental masses during the Late Paleozoic and Early Mesozoic Eras.

Panthalassa— The enormous ocean that surrounded Pangaea.

Pascal— A unit used to measure pressure.

peak ring— A ring of mountainous peaks, usually composed of crystalline basement rocks, which forms around the inner basin of a complex impact crater.

pebble— A rock clast ranging in size from 0.16 inch to 2.5 inches.

phenacodont— One of a group of primitive, sheep-sized, hoofed mammals characteristic of the Eocene Epoch.

plankton— Organisms that drift, usually in the upper parts of a water body, especially lakes and oceans.

Pleistocene— The most recent of Earth's several ice ages, lasting from 1.8 million to 10 thousand years ago.

pollen— A reproductive element of flowering plants. Fossilized pollen is used to date rock formations, especially non-marine rocks, and to interpret past climatic conditions.

radiolarian— One of a group of microscopic animals, which construct lacy shells of silica. Radiolarian fossils are widely used to date rock formations and to interpret their depositional environments.

relative sea level— The height of sea level as compared to some "stationary" object on Earth; but because the Earth's crust also can move up or down, no object can be considered stationary. Therefore, all such sea-level measurements are relative.

reverse osmosis— The process whereby pressure is applied to force water through a semipermeable membrane from a salt-rich solution into a salt-poor solution. Natural osmosis causes water to flow from the salt-poor solution into the salt-rich solution.

scarp— A steep slope, often associated with a fault or some other abrupt geological or geographic barrier.

schist— A medium- to coarse-grained metamorphic rock containing mainly plate-like minerals, which are oriented in parallel thin sheets.

sedimentary rock— A rock unit composed of particles or fragments eroded from other rocks and usually deposited by wind, water, or gravity flow.

seismic profile — A two-dimensional cross section of subsurface rock formations produced by a seismic reflection survey.

seismic reflection — A dark band on a seismic profile produced by sound energy reflected from the boundary between two layers of rock that have different compositions or densities.

shocked quartz — A quartz grain whose internal crystalline structure has been disrupted by the enormous force of a meteorite impact.

spore — The microscopic one-celled reproductive body of a plant, often adapted to survive harsh environmental conditions. Spores often are the only fossils preserved in terrestrial rocks. Spores are widely used to date rock formations and to determine their environments of deposition.

stishovite — One of the high-temperature forms of silica produced by the impact of a meteorite.

streamer — A long plastic hose, filled with insulating oil, which houses a series of hydrophones; it is towed behind the ship during a marine seismic reflection survey.

stromatolite — A mound or stony edifice built of stacked microbial mats. Stromatolites are the main large fossils of the Proterozoic Era.

taeniodont — One of a group of bear-sized, plant-eating Eocene mammals.

taxa — Members of any formally named category of organisms, ranging from species to kingdom; the singular form is *taxon*.

tektite — A fragment of melt-glass formed from the Earth's crust by the heat of a meteorite impact.

test — A technical term for "shell," derived from the Latin word for shell (*testa*).

Tethys — An ancient ocean that encompassed Earth's equatorial belt during the Eocene and other epochs of the Cenozoic Era.

tetrapod — A term of Greek derivation, used to designate four-footed animals.

thruster — A propeller used in large ships, such as the *Glomar Challenger*, which is directed sideways to control lateral position.

tillodont — One of a group of large, bear-sized, plant-eating Eocene mammals.

Triassic — The period of geologic time from 245 to 185 million years ago, in which reptiles began to dominate terrestrial and marine habitats.

trilobite — One of a group of joint-legged benthic marine arthropods, whose three-lobed body fossils are extremely abundant in rocks of the early Paleozoic Era.

tsunami — A Japanese word originally used to describe large waves produced by earthquakes, but now used also for meteorite-generated waves.

Suggestions for Further Reading

Alvarez, Walter. *T. rex and the Crater of Doom.* Princeton: Princeton University Press. 1997.

Barnes-Svarney, Patricia. *Asteroid: Earth Destroyer or New Frontier?* New York: Plenum Press. 1996.

Chapman, Clark R., and Morrison, David. *Cosmic Catastrophes.* New York: Plenum Press. 1989.

Cox, Donald W., and Chestek, James H. *Doomsday Asteroid: Can We Survive?* New York: Prometheus Books. 1996.

Desonie, Dana; Levy, David H.; and Shoemaker, Eugene. *Cosmic Collisions.* New York: Henry Holt. 1996.

Emiliani, Cesare. *Planet Earth—Cosmology, Geology, and the Evolution of Life and Environment.* Cambridge: Cambridge University Press. 1992.

Erwin, Douglas H. *The Great Paleozoic Crisis—Life and Death in the Permian.* New York: Columbia University Press. 1993.

Fortey, Richard. *Life: A Natural History of the First Four Billion Years of Earth History.* New York: Alfred A. Knopf. 1998.

Glen, William. *The Mass Extinction Debates: How Science Works in a Crisis.* Stanford: Stanford University Press. 1994.

Gould, Stephen Jay. *Wonderful Life.* New York: W. W. Norton & Company. 1989.

———, ed. *The Book of Life.* New York: W. W. Norton & Company. 1993.

Hodge, Paul. *Meteorite Craters and Impact Structures of the Earth.* Cambridge: Cambridge University Press. 1994.

Hsü, Kenneth J. *The Great Dying.* New York: Ballantine Books. 1986.

———. *Challenger at Sea—A Ship That Revolutionized Earth Science.* Princeton: Princeton University Press. 1992.

Levy, David H. *Impact Jupiter: The Crash of Comet Shoemaker-Levy 9.* New York: Plenum Press. 1995.

Lewis, John S. *Rain of Iron and Ice—The Very Real Threat of Comet and Asteroid Bombardment.* Reading: Addison-Wesley, Helix Books. 1996.

Norton, O. Richard; Marvin, Ursula B.; and Norton, Dorothy S. *Rocks from Space: Meteorites and Meteorite Hunters.* Missoula: Mountain Press, 2d ed. 1998.

Prothero, Donald R. *The Eocene-Oligocene Transition: Paradise Lost.* New York: Columbia University Press. 1994.

Spencer, John R.; Mitton, Jacqueline; and Spencer, John G. *The Great Comet Crash: The Impact of Comet Shoemaker-Levy 9 on Jupiter.* Cambridge: Cambridge University Press. 1995.

Vogt, Gregory L. *The Search for the Killer Asteroid.* New York: Millbrook Press. 1994.

Ward, Peter Douglas. *The End of Evolution: A Journey in Search of Clues to the Third Mass Extinction Facing Planet Earth.* New York: Bantam Books. 1995.

Recommended Sites on the World Wide Web

Asteroid and Comet Impact Hazards — http://impact.arc.nasa.gov/index.html
http://impact.skypub.com/page1.html

Evolution and Extinction — http://www.bbc.co.uk/education/darwin/

Jet Propulsion Laboratory — http://www.jpl.nasa.gov

Lunar and Planetary Institute — http://cass.jsc.nasa.gov/meetings/lpsc97/
lpi.html

Mass Extinctions — http://www.wf.carleton.ca/Museum/extinction/
homepg.html

Meteorites, Asteroids — http://inferno.physics.uiowa.edu/˜jdf/lec40/
lec40.html

Planetary Data System — http://pds.jpl.nasa.gov/

Planetary Geology — http://europa.la.asu.edu/def/html/

Spacewatch Program — http://www.impact.arc.nasa.gov/index.html

Spacewatch Project — http://www.lpl.arizona.edu/spacewatch/

Terrestrial Impact Structures — http://gdinfo.agg.emr.ca/toc.html?/crater/
world_craters.html

3-D Tour of Solar System — http://cass.jsc.nasa.gov/research/stereo_atlas/
SS3D.html/

Tour of Universe — http://hoku.ifa.hawaii.edu/astrokid/files/tour.html

USGS Space Science and Maps — http://wwwflag.wr.usgs.gov/USGSFlag/
Space/space.html

Views of Solar System — http://www.hawastsoc.org/solar/

Index

Italicized page numbers refer to figures and tables.

Accomack, county of, 15
Act: Clean Water, 144; Comprehensive Environmental Response, Compensation, and Recovery, 144; National Environmental Policy, 144; Resource Conservation and Recovery, 144; Safe Drinking Water, 144; Toxic Substance Control, 144
Africa, 132; South, 61, 141
Age: of Bacteria, 70, 78; of Chemistry, 68; of Fishes, 75; of Insects, 78; of Mammals, 73, 78
Agena, Warren, 14
air gun, 12, 13, 128, 130; definition of, 157; GI, 131. See also hydrophones
Albemarle Sound, 22
Aldrich, Tom, 132
algae: of Archaean Era, 69, 72; of Cambrian continents, 72; first appearance of, 69; known species of, 83; marine, 26, 71; primitive, 73
Ames: crater, 141; impact structure, 141; National Laboratory, 44; Ridge, 109, 110
ammonia, in Earth's primordial atmosphere, 68
ammonites: extinction of, 78; of Mesozoic Era, 73, 77
amphibians, 76, 77; archaic, 76; of Carboniferous Period, 76; of Devonian Period, 69; extinction of, 85; first appearance of, 69; of Phylum Craniata, 83; primitive, 73
Anabar shield, 98
angiosperm: definition of, 157; evolution of, 78; first appearance of, 77
Animalia, Kingdom of, 84
annular trough, 38; of Chesapeake Bay crater, 53, 55, 64; of complex crater, 39, 40; definition of, 157; of Manicouagan crater, 138; of Toms Canyon crater, 43

Anomalocaris, 73; in Burgess Shale, 75
Antarctica: glaciation of, 103; ice sheets of, 91, 93, 102, 106; meteorites of, 149, 151; ozone hole over, 70
Appalachian Mountains, 3-5, 119; flanks of, 106; foothills of, 50; geochemistry of, 36; rocks of, 5, 36
aquifer, 6-8, 15, 121; contamination of, 116, 143; definition of, 157; of Delmarva Peninsula, 116; above Exmore breccia, 115, 116, 146; load pressure on, 121; of Mattaponi Formation, 10; saltwater, 144; in tectonic movement, 121; temperature of, 121. See also confining unit
Archaean Era, 69, 70-72. See also Era
Archaeotherium, 92, 95
arthropod, 72, 76, 162; ancestral, 74
ash, from impact-generated wildfires, 97
Associated Press, 65
asteroid, 11, 57, 59; belt of, 148, 149; cluster of, 147; collisions of, 149; danger from, 149, 157; definition of, 147; impact of, 153; iridium in, 81; size of, 149; tracking of, 153; Vesta, 149; 1997 XF11, 153
Atlantic: Ocean, 5, 11, 16, 51, 55, 63, 77, 83, 101, 105-8, 111, 116, 119, 123, 145; City, 24, 31, 44, 51; coastal plain of, 110, 131; continental margin of, 50, 115; continental shelf of, 28; Eocene shoreline of, 4, 95; during ice age, 114; Marine Geology, 8; seabed of, 23, 96; seaboard of, 41; seaway of, 105, 106, 118
atmosphere: blast wave in, 55; carbon dioxide in, 68, 85, 93, 98; chlorine in, 97; debris in, 55; hydrogen in, 68; hydrogen sulfide in, 68; hypercane in, 96; of Jupiter, 148; laser beam in, 154; meteorites in, 55, 151; methane in, 68; ozone in, 70, 97; oxygen in, 70, 85; pressure of,